Foraging Strategies and Natural Diet of Monkeys, Apes and Humans

Foraging Strategies and Natural Diet of Monkeys, Apes and Humans

Proceedings of a Royal Society Discussion Meeting held on 30 and 31 May 1991

Organized and edited by
A. Whiten and E. M. Widdowson

CLARENDON PRESS · OXFORD
1992

Oxford University Press, Walton Street, Oxford OX2 6DP

Oxford New York Toronto
Delhi Bombay Calcutta Madras Karachi
Petaling Jaya Singapore Hong Kong Tokyo
Nairobi Dar es Salaam Cape Town
Melbourne Auckland

and associated companies in
Berlin Ibadan

Oxford is a trade mark of Oxford University Press

Published in the United States
by Oxford University Press, New York

A catalogue record for this book is available from the British Library

Library of Congress Cataloging in Publication Data
(Data available)

ISBN 0–19–852255–X

Printed in Great Britain
by Cambridge University Press

Preface

The emergence of techniques of husbandry 10000–15000 years ago marks off a relatively tiny segment in the approximately two-million-year-old history of the genus *Homo*. For most of this longer period, hominid diet was 'natural' in the sense that it was harvested as a natural resource from what the environment of the moment had to offer. This suggests a long-term process of adaptation between on the one hand, foraging strategies and dietary preferences and, on the other, particular arrays of natural foods and their dietary constituents which characterized evolving hominids' habitats.

The papers collected in this volume seek further understanding of the nature and evolution of these diets and their precursors. To achieve this we have brought together new studies being undertaken within a number of related scientific disciplines. First come studies of foraging and dietary strategies in non-human primates, against which the emergence of hominid feeding patterns must be understood. A second group of papers attempts to reconstruct the principal features of proto-hominid and hominid dietary evolution. A third section turns to a different source of information, the diets of present-day peoples who have continued to gather and hunt natural foods. Fourthly and finally, we attempt to complete the picture by tracing major changes and variability, or lack of it, in diet which the greater number of human societies have experienced in more modern times.

We are not the first to engineer such a mix of perspectives on the topic of diet in human evolution. Ten years ago, Robert Harding and Geza Teleki published *Omnivorous primates*. Under the subtitle *Gathering and hunting in human evolution*, they likewise drew together contributions from primatology, archaeology, and hunter–gatherer research. These incorporated many new findings based on a growing tendency towards systematic and quantitative assessments of foods and feeding.

However, Harding was forced to conclude his comprehensive review of feeding behaviour in 129 species of primate with the admission that it was still not possible 'to confront some of the most fundamental issues in primate dietary research today: What are the nutritive requirements of each species? What is the nutritive content of each food item eaten? And, by extension, what are the exact quantities of each food consumed? With few exceptions such data are not available; yet it is in precisely these areas that research must be undertaken if we are to come to a full understanding of diets and the role they play in the evolutionary development of primate anatomy and behaviour. Such research is complex, costly, and difficult, and will require the participation of many different branches of science if it is to be successful... Not until a basis of quantitative data has been collected from this kind of research can evolutionary hypotheses tracing the ultimate origin of mankind to changing diets be much more than speculative.' (p. 200)

In the 1980s we have seen significant progress in the recording and reporting of these crucial but elusive details of nutritional information for both wild-living non-human primates and for human hunter–gatherers. This achievement underlies the results reported in the first and third sections of this volume.

After an overview of non-human primate diets and nutritional requirements, studies focusing on two particular taxa are included. Two examine the feeding behaviour of those species of ape most closely related to us: chimpanzees and gorillas. As it is with African apes we shared our last common ancestor among non-human primates, and as they appear to have remained in habitats more akin to those of earlier apes, they provide important insights into the nature of proto-hominid diets.

Baboons merit special attention for a different reason. Of interest here are the ways in which a medium-sized anthropoid primate adapts its foraging habits to the distribution and nature of foodstuffs in savannah ecologies, which are similar to those exploited during important phases of hominid evolution.

Modern-day chimpanzees, gorillas, and baboons, however, cannot be taken as literal models of ancestral states. The role of comparative research is rather to reveal general rules of dietary adaptation, to which our ancestors would also have been subject. Significant intraspecific variations in the feeding patterns of monkeys and apes have also emphasized how important it is to study several different populations of the same species before attempting generalizations. If this is the case for non-human primates it is certainly no less so for humans. In drawing together recent studies of the diets and foraging methods of tropical and sub-tropical human hunter–gatherers we have deliberately included research from Africa, South America, and Australia.

Numerous factors conspire to make these hunter–gatherers imperfect models of ancestral foragers. The people themselves are of course fully modern: they are not living fossils any more than are the modern baboons and chimpanzees studied in the first section. What is important is that they are operating under ecological constraints that resemble those of earlier times. This is imperfectly the case for such reasons as contact with neighbouring farming peoples and restriction to marginal habitats: further arguments to examine as broad a sample of peoples as possible.

Information about these diets becomes particular valuable when examined in conjunction with the archaeological record. The relationship here is complementary: archaeology provides an essential framework within which data from living species can be sensibly interpreted, but the archaeological record can tell us rather little about certain aspects of foraging which by contrast can be minutely detailed for living species.

The second section is therefore designed to describe the transition from dietary patterns last shared with other primates as exemplified in the first section, to established human hunter–gatherer diets such as are later described in the third section.

The past decade has seen remarkable ingenuity in the development of archaeological techniques that extend the inferences that can be drawn from preserved materials. These include microscopic examination of the surfaces of teeth, stone tools and prey bones to draw dietary distinctions that can be checked against those demonstrable in modern species (in the case of teeth) or experimentally re-created (in the case of tool usage and effects on prey bones). Taphonomy, the study of the effects of the fossilization process itself, has become a significant subdiscipline in its own right. This has added further rigour to the endeavour of making inferences about the past, as has the experimental testing of hypotheses on realistic modern materials.

Into the fourth and final section of the volume are compressed accounts of human dietary change and variation through the development of agriculture up to the present day. The nature and variability of the range of human foods are assessed. Major health and nutritional problems from early neolithic times to the present day indicate the legacy of our natural dietary adaptations and show how far we still are from being able to apply nutritional knowledge to the well-being of the human species.

We thank the chairmen who so skilfully facilitated the productivity of the Royal Society Discussion Meeting: Professor Stuart Altmann, Professor Bernard Wood, and Professor John Waterlow.

Special thanks are also due to Peter Warren, Christine Johnson, and Simon Gribbin at the Royal Society for their warm and efficient support.

A.W.
E.M.W.

Contents

Human diets: prehistory to present day

The nutritional consequences of foraging in primates: the relationship of nutrient intakes to nutrient requirements

OLAV T. OFTEDAL

Department of Zoological Research, National Zoological Park, Smithsonian Institution, Washington, D.C. 20008, U.S.A.

SUMMARY

Many studies have examined the proportion of time that primates devote to feeding on various types of food, but relatively little is known about the intake rates associated with each food. However, the nutritional consequences of foraging can only be interpreted by comparing nutrient intakes with estimated nutrient requirements. The energy available to primates from ingested foods will depend both on the composition of the food and the extent to which various constituents, including fibre fractions, are digested. Both human and non-human primates have relatively low requirements for protein as a consequence of slow growth rates, small milk yields and relatively dilute milk. Because the nutrient demands of growth and reproduction are spread out over time, it appears that primates do not need to seek out foods of particularly high nutrient density, except perhaps during weaning. Although food selection in some species of primates appears to be correlated with the protein concentration of foods, it is unlikely that high dietary protein levels are required, at least when foods of balanced amino acid composition (such as leaves) are included in the diet.

1. INTRODUCTION

Most primates are omnivores and consume a complex variety of foods. In a review of field studies of the diets of 131 primate species, Harding (1981) concluded that fruit was consumed by 90% of the species, soft plant foods (immature leaves, buds, shoots, flowers) by 79%, mature leaves by 69%, invertebrates by 65%, seeds by 41% and other animal foods (including eggs) by 37%. Yet despite the wide variety of foods eaten, the particular items selected represent but a fraction of the 'potential foods' available. The causal factors that have moulded the patterns of food selection and avoidance are undoubtedly diverse. Food choice may be influenced by energy needs, requirements for specific nutrients, constraints of the digestive system, digestion-inhibiting or toxic constituents in foods, difficulties in removing inedible components, degree of food clumping and dispersal, intra- and interspecific competition, and predation (see, for example, Milton (1980, 1984); McKey *et al.* (1981); Glander (1982); Altmann *et al.* (1987); Janson (1988)).

Regardless of the factors underlying food choice, foraging behaviour can be considered successful only if the diet obtained provides sufficient amounts of energy and other nutrients to fulfil the nutritional requirements of the forager. However, this seemingly simple criterion is hard to evaluate for free-ranging primates. In this paper I shall discuss some of the complexities in evaluating the nutritional consequences of foraging patterns in primates.

2. THE PROPORTIONAL CONTRIBUTION OF DIFFERENT FOODS

The seasonal pattern of food consumption by primates is usually determined by direct observation, unless dense vegetation so limits observation that other methods become necessary (see Calvert 1985). For example, in a study of feeding by red howler monkeys (*Alouatta seniculus*) in the llanos of Venezuela, we found that mature leaves accounted for 30% of the time spent feeding on major foods in the dry season, increasing to 67% in the wet season (M. S. Edwards, S. D. Crissey, O. T. Oftedal & R. Rudran, unpublished data). Flowers and young leaves were important in the dry season (31 and 14%, respectively) but not in the wet season (0 and 3%), whereas fruits were equally important in both the dry (25%) and wet (30%) seasons.

Although botanical categories are appropriate for ecological description of feeding behaviour, they may have little value in predicting nutritional composition. For example, the common notion that fruits are lower in fibre fractions and thus of 'higher quality' than leaves does not always hold (table 1; Calvert 1985; Barton *et al.* 1992). On a dry-matter basis the mean concentrations of neutral-detergent fibre (NDF), acid-detergent fibre (ADF), lignin and several minerals were remarkably similar among the various food categories consumed by red howlers (table 1). Moreover large variation occurred among foods in each category, especially with respect to protein, lignin, and calcium.

Phil. Trans. R. Soc. Lond. B (1991) **334**, 161–170
Printed in Great Britain

Table 1. *Percentage composition of the major food plants eaten by red howlers*[a]

	flowers ($n = 3$)	fruit ($n = 9$)	mature leaves ($n = 20$)	young leaves ($n = 5$)
dry matter	25.1±4.86	23.7±1.07	36.5±2.23	32.2±2.28
water	74.9±4.86	76.3±1.07	63.5±2.23	67.8±2.28
protein	14.4±3.06	7.0±1.12	16.6±1.25	21.2±3.76
fibre (NDF)	50.6±3.22	53.8±5.00	57.2±1.89	54.4±5.64
fibre (ADF)	35.8±3.22	35.2±4.55	40.5±1.91	36.4±6.28
acid lignin	17.1±3.51	16.6±2.43	20.4±1.99	21.1±5.34
calcium	0.49±0.22	0.64±0.17	1.36±0.22	0.29±0.07
phosphorus	0.30±0.01	0.16±0.05	0.14±0.01	0.28±0.06

[a] Major food plants are defined as those that collectively comprise 80% of observed feeding time in a particular month (M. S. Edwards, S. D. Crissey, O. T. Oftedal & R. Rudran, unpublished data). Data presented as mean±s.e.m. Young leaves were distinguished from mature leaves on the basis of size, shape, colour and texture. NDF = neutral-detergent fibre, ADF = acid-detergent fibre (see Van Soest 1982); n = number of species, protein = (total nitrogen) × 6.25.

Table 2. *Examples of calculated ingestion rates of adult male red howlers at Hato Masaguaral, Venezuela*[a]

part...	flowers	ripe fruit	simple leaves	compound leaves
species...	*Pithecellobium saman*	*Mangifera indica*	*Ficus pertusa*	*Pithecellobium tortum*
rank[b]...	DS 1	DS 2	WS 1	WS 4
food intake[c]				
bite rate/(bite min^{-1})	14.3	7.8	12.6	6.4
bite size/(g per bite)	0.75	2.39	0.45	0.39
intake/(g min^{-1})	10.7	18.7	5.7	2.5
food composition				
dry matter (DM) (%)	20.2	20.3	30.2	39.2
protein (% DM)	19.8	4.1	12.2	22.0
fibre (NDF) (%DM)	47.4	21.0	49.1	59.2
calcium (% DM)	0.16	0.11	3.06	0.52
phosphorus (% DM)	0.30	0.04	0.08	0.08
nutrient intake				
dry matter (g min^{-1})	2.17	3.81	1.71	0.97
protein (g min^{-1})	0.43	0.16	0.21	0.21
fibre (NDF) (g min^{-1})	1.02	0.80	0.84	0.57
calcium (mg min^{-1})	3.5	4.2	52.3	5.0
phosphorus (mg min^{-1})	6.5	1.5	1.4	0.8

[a] Unpublished data of M. S. Edwards, S. D. Crissey, O. T. Oftedal and R. Rudran.
[b] Rank refers to relative importance in terms of percent of feeding time in dry (DS) and wet (WS) seasons: 1 = most important, 2 = second most important, etc.
[c] Bite rate data from field observations; bite size data from measurements on captive adult males at Hato Masaguaral (10–14 intake trials per plant part, greater than 10 bites per intake trial).

Within a plant species immature leaves are usually lower in fibre than are mature leaves (Milton 1979; McKey *et al.* 1981). However many primates appear to select leaves that are low in fibre fractions, regardless of the growth stage of the leaves (for examples, see Milton (1979); Oates *et al.* (1980); Glander (1981); McKey *et al.* (1981); Calvert (1985)). Thus there may be little difference in the fibre concentrations of the young and mature leaves actually ingested (table 1; Glander 1981).

3. MEASUREMENT OF FOOD INTAKE

Data on time spent feeding do not take into account the considerable differences in intake rate that may occur during feeding on different foods (Hladik 1977; Milton 1984). Food intake rates certainly vary in red howlers (table 2): in 1 min of feeding on large, ripe mangoes (*Mangifera indica*), red howlers ingest as much food (by mass) as when feeding on compound leaves of *Pithecellobium tortum* for 7.5 min. The high intake rate when feeding on ripe mangoes reflects the large bite size (2.4 g per bite). Flowers of the saman tree (*Pithecellobium saman*) are both large and quickly consumed (14 bites per minute). The twofold difference in intake rate of the two leaf types is due to the difference in bite rate (12.6 versus 6.4 bites per minute), not bite size.

Nutrient intake rates can be calculated from data on food intake and composition (table 2). Dry matter intake rate was highest for mango fruit, protein and phosphorus intake rates were greatest for saman flowers, and calcium intake rate was highest for fig (*Ficus pertusa*) leaves (table 2). Fibre intake rates were lowest when the monkeys were feeding on *Pithecellobium* leaves. These examples illustrate that maximum intake rate of a nutrient does not always occur when a primate feeds on the food item with the highest concentration of the nutrient. For example, red howlers feeding on fig leaves containing 12% protein achieved the same protein intake rate as when feeding on leguminous leaves containing 22% protein, although the howlers simultaneously ingested more dry matter, one third of which was NDF (table 2). The potential cost or benefit of the additional dry matter and fibre consumed depend on the digestibility of this material, whether it contains potentially toxic compounds, and whether the indigestible component exerts a negative, bulk-limiting effect on overall food consumption (Van Soest 1982).

4. THE ENERGY AVAILABLE FROM PRIMATE FOODS

Nutrient intake rates need to be related to the intake of metabolizable energy (ME), as maintenance of energy balance is the ultimate determinant of food intake (National Research Council 1978). Although total gross energy is easily measured in plant materials by bomb calorimetry, the proportion of this energy available to the animal may be highly variable, depending to a large extent on the amount, type and fermentability of fibre. Fibre fractions are resistant to mammalian digestive enzymes, but may be fermented by symbiotic microorganisms in the digestive tract, leading to net production of volatile fatty acids. Volatile fatty acids are undoubtedly an important energetic substrate for many primates, especially those species with specialized fore- or hind-gut fermentation areas (Bauchop & Martucci 1968; Milton & McBee 1983; Martin *et al.* 1985).

The fibre in plant materials is chemically complex, often including some readily fermentable carbohydrate (e.g. pectin), partially fermentable structural carbohydrate (e.g. cellulose and hemicellulose) and polyphenolic compounds that are thought to be completely indigestible (e.g. lignin). In the detergent fibre method of analysis, the NDF fraction represents the entire plant cell wall (other than pectin and some minor components that are solubilized in neutral detergent), whereas the ADF fraction represents cellulose and lignin (Van Soest 1982). Both fractions can constitute a large proportion of primate foods (table 1), making measurement of fibre digestibility an essential step in the assessment of metabolizable energy.

In captivity, folivorous or omnivorous primates consuming diets of relatively low fibre concentration (15–25% NDF, dry matter basis (DMB)) have been shown to be able to digest a large proportion of the fibre. For example, black and white colobus (*Colobus guereza*) digested 68–81% of NDF and 68–69% of ADF, whereas chimpanzees (*Pan troglodytes*) digested 71% of NDF and 57% of ADF (Oftedal *et al.* 1982; Watkins *et al.* 1985; Milton & Demment 1988). On higher fibre diets, digestibility of fibre fractions may be reduced, apparently due to more rapid passage of digesta through the tract. NDF and ADF digestibilities were only 54% and 33%, respectively, when chimpanzees were fed a manufactured diet containing 34% NDF (Milton & Demment 1988).

The manufacture of dry primate feeds involves mixing of ground ingredients, introduction of steam, and expression through an extruder under pressure. In the digestive tract, these feeds disintegrate into small particles, providing large surface area for attachment of microbes, thereby facilitating fermentation. The difference in physical form and the higher fibre concentration of many natural foods suggests that primates in the wild may not be able to digest fibre nearly as completely as the above experimental values suggest. Milton *et al.* (1980) estimated that mantled howlers (*Alouatta palliata*) digested only 24–42% of NDF when they were fed diets comprised of mixtures of natural foods (leaves and fruit).

Because human foods are typically low in fibre and high in digestibility, it is common in the United States to apply energetic factors of 4 kcal (16.7 kJ) per g protein and carbohydrate and 9 kcal (37.7 kJ) per g fat in estimating physiologically available energy in human diets (National Research Council 1989). In this system 'carbohydrate' percentage is estimated by difference (100 − (% water + % fat + % protein + % ash)) and hence includes fibre fractions. This system overlooks differences in the digestibility and fibre concentration of foods and is inappropriate for the natural foods eaten by primates. Even flowers and fruits may be quite high in fibre fractions (table 1). Insects also contain significant but variable amounts (8–27% of dry matter) of the relatively indigestible structural carbohydrate, chitin (Allen 1989).

Ideally, digestion trials should be conducted with each primate food to measure the amount of energy that can be extracted via digestive processes. We have been able to capture troops of free-ranging red howlers and adapt them to long-term studies in specially constructed cages in the field (Crissey *et al.* 1991; see also Milton (1980)). Unfortunately digestion trials are laborious and impractical for foods that are difficult to collect in the large amounts required for the trials. *In vitro* fermentation assays with rumen innocula or enzymic procedures that employ proteases and cellulases have been used to generate theoretical indices of digestibility for plant materials eaten by ruminants (Barnes 1973; Van Soest 1982). These procedures have also been applied to plants ingested by primates (see, for example, Oates *et al.* (1980); McKey *et al.* (1981); Calvert (1985)), with the implicit assumption that folivorous primates with fore- or hind-gut fermentation will digest foods in a similar fashion and to the same extent as a ruminant. However *in vitro* assays do not take into account important differences in tooth structure and function, gut morphology, rates of digesta passage, particle size segregation and other aspects of the digestive process. It is essential that *in vitro* indices be validated by digestion trials with the primate species of interest.

Any particular food will not be digested equally well by all primates. For example, Power (1991) has shown that the energy digestibility of an artificial diet may range from 71% to 86% among different species of

Table 3. *Comparison of recommended nutrient levels (on a dry matter basis) for human and non-human primates when consuming equivalent diets*[a]

	human			non-human primates (all stages)
	infant	young woman	lactating woman	
age/years	0.5–1	19–24	1st 6 months	—
energy/(kcal d^{-1}) intake	850	2200	2700	—
dry matter/(g d^{-1}) intake	179	550	675	—
protein (%)	7.8	8.4	9.6	16.3
calcium (%)	0.34	0.22	0.18	0.54
phosphorus (%)	0.28	0.22	0.18	0.43
magnesium (%)	0.034	0.051	0.053	0.16
iron (p.p.m.)	56	27	22	196
zinc (p.p.m.)	28	22	28	11
iodine (p.p.m.)	0.28	0.27	0.30	2.2
selenium (p.p.m.)	0.08	0.10	0.11	—
vit. A/(IU kg^{-1})	6985	4848	6420	10900
vit. D/(IU kg^{-1})	2235	727	593	2170
vit. E/(IU kg^{-1})	22	15	18	54
vit. K (p.p.b.)	56	109	96	—
vit. C (p.p.m.)	196	109	141	109
thiamin (p.p.m.)	2.2	2.0	2.4	—
riboflavin (p.p.m.)	2.8	2.4	2.7	5.4
niacin (p.p.m.)	33.5	27.3	29.6	54
vit. B_6 (p.p.m.)	3.4	2.9	3.1	2.7
folate (p.p.b.)	196	327	415	217
vit. B_{12} (p.p.b.)	2.8	3.6	3.9	—

[a] Calculated from National Research Council (1978, 1989), based on the following assumptions.
1. Manufactured primate diets contain 92% dry matter, so primate requirements as given in National Research Council (1978) have been divided by 0.92.
2. A human diet equivalent to a manufactured primate diet would contain 4 kcal per g dry matter, so energy allowances have been divided by 4 (4.75 for infant fed some milk) to calculate equivalent dry matter intakes.
3. Bioavailability of nutrients in mixed human diets and primate diets are comparable, so no adjustments in nutrient levels are necessary.

callitrichids (marmosets and tamarins). With an interspecific decline in body mass from 680 to 310 g, callitrichids exhibit a decrease in the transit time of marked food through the digestive tract and a corresponding decrease in energy digestibility. However the smallest callitrichid, the pygmy marmoset (*Cebuella pygmaea*), appears to be an exception to this pattern, for transit time is relatively long and energy digestibility is high (84%). This difference in the digestive function of pygmy marmosets may reflect specialization on gums that require fermentation (Power 1991), and illustrates the importance of species-specific data on the digestibility of foods.

5. PRIMATE NUTRIENT REQUIREMENTS

The relative advantage or disadvantage associated with a high or low intake rate of a particular nutrient will depend on the total intake and bioavailability of the nutrient in relation to requirements of the animal. There may be a significant premium in terms of survival and ultimate reproductive success for a primate that is able to augment the intake of a limiting or deficient nutrient, but little if any benefit in increasing the intake of a nutrient that is already present in sufficient amounts to meet requirements.

It has been shown that laboratory and farm animals require about 45–47 nutrients (vitamins, minerals, amino acids, fatty acids), depending on the species. Although the Panel on Nonhuman Primate Nutrition of the Committee on Animal Nutrition, U.S. National Research Council was only able to estimate quantitative requirements for 24 nutrients (National Research Council 1978), there is little doubt that primates require most if not all the nutrients known to be required by other mammals.

The estimated nutrient requirements of non-human primates may be compared with recommended nutrient levels for humans by expressing both sets of estimates on a dry matter basis (table 3). The human recommendations are not minimal requirements since they include allowances for the bioavailability of nutrients in typical diets, and have been increased to encompass expected variability among individuals. Even with these adjustments, the human recommendations for most nutrients are lower than the estimated requirements of non-human primates. Most of the non-human primate requirements are not well defined, leading to a reliance on 'practical levels' as best estimates of requirements. 'Practical levels' are those thought to be sufficient based on experience with primate colonies, but may be considerably higher than minimal requirements.

In formulation of manufactured feeds it is usually safer to err on the high side for a given nutrient, adding a margin of safety to cover losses during manufacture

Table 4. *Comparison of nutrient composition of commercial primate diets to nutrient requirements of non-human primates as estimated by the U.S. National Research Council (1978)*[a]

	primate requirement[b]	commercial primate diets		
		mean	range	*n*
dry matter (dry diets) (%)	—	91.7	89.2–93.6	5
dry matter (canned diets) (%)	—	41.7	41.1–42.3	2
fat (%)	—	5.5	2.7–9.8	7
protein[c] (%)	16.3	21.2	16.0–26.1	7
neutral detergent fibre (%)	—	20.9	14.0–25.7	6
acid detergent fibre (%)	—	8.0	4.9–14.0	6
acid lignin (%)	—	1.7	0.8–2.4	6
ash (%)	—	6.8	5.6–8.3	7
gross energy/(kcal g^{-1})	—	4.63	4.43–4.89	5
metabolizable energy[d]/(kcal g^{-1})	—	3.91	3.49–4.24	7
calcium (%)	0.54	1.29	0.98–1.84	7
phosphorus (%)	0.43	0.66	0.40–0.92	7
magnesium (%)	0.16	0.18	0.10–0.29	7
sodium (%)	—	0.40	0.19–0.71	7
potassium (%)	—	1.05	0.83–1.44	7
iron (p.p.m.)	196	492	111–1140	7
copper (p.p.m.)	—	17.6	14.0–22.6	7
zinc (p.p.m.)	11	196	106–505	7
manganese (p.p.m.)	—	89	31–176	7
selenium (p.p.m.)	—	0.36	0.07–0.59	7

[a] All values (except dry matter), expressed on a dry matter basis. Protein = total nitrogen × 6.25. Analytical data on commercial primate diets provided by Dr Mary E. Allen, Allen and Baer Associates, Olney, Maryland 20832.
[b] Requirements for primates converted to a dry matter basis assuming an average dry matter concentration of natural ingredient diets of 92%.
[c] The National Research Council (1978) notes that New World primates may require up to 27% protein (DM basis). Commercial diets marketed for New World primates usually have 25–26% protein.
[d] Metabolizable energy (ME) of diet calculated assuming ME values of 4 for protein and carbohydrate (NFE) and 9 for fat.

and storage as well as possible interspecific differences in requirements. Analytical data for commercial primate feeds illustrate that some nutrients are included at higher levels than the NRC estimated requirements (table 4). These feeds also vary considerably in composition, in part because some feeds have been targeted to specific types of primates (e.g. callitrichids, cebids or folivorous species) and also because the ingredients and mineral premixes used by various manufacturers are different.

It is not possible in this paper to review the importance, function and method of establishing requirements for all nutrients that are considered essential for primates. However, in evaluating compositional data on foods consumed by free-ranging primates, one must have some basis for judging the level of a nutrient to be high or low. The values in tables 3 and 4 provide benchmarks that may be used for comparison. Because food intake tends to decline with increasing energy density, somewhat higher nutrient concentrations may be appropriate for diets that are high in fat and energy. Conversely, the lower ME concentration of high-fibre diets may permit reduced nutrient levels without compromising nutrient intakes. It is important to recognize that the bioavailability of nutrients may differ greatly among foods. For example, compounds such as phytic acid, oxalic acid, protease inhibitors and tannins may reduce digestibility or bioavailability of particular minerals or protein (Swain 1979; Van Soest 1982; Cheeke & Shull 1985; Morris 1986). High levels of some minerals may also produce toxicity or interfere with the utilization of other minerals (National Research Council 1980).

6. ESTIMATES OF THE PROTEIN REQUIREMENTS OF PRIMATES

Protein has frequently been singled out as potentially beneficial to foraging primates (see, for example, Hladik 1977; Milton 1979, 1981; Glander 1981; McKey *et al.* 1981; Calvert 1985; Altman *et al.* 1987; Whiten *et al.*, this symposium). It has been suggested that primates exhibit preference for foods high in protein and essential amino acids, whether in selecting young leaves rather than mature leaves, in making choices among different species of leaves, fruits or seeds, or in supplementing fruit diets with a protein source such as leaves or insects.

The attention to protein status was probably inspired by the widespread view in the 1960s that human protein malnutrition was 'the major nutritional problem of the world', and that protein deficiency was 'common' in the less developed countries where primate field studies are usually undertaken (FAO 1965). Subsequently a more balanced view emerged in assessments of human nutrition, recognizing the predominant role of inadequate energy intake in protein–energy malnutrition (PEM) (Widdowson, this symposium). Even when protein intakes are low, clinical symptoms usually emerge in association with infection

Table 5. *Estimated protein requirements of primates*

species	data source[a]	age class	body mass (kg)	type of protein	principal parameter	estimated protein requirements: intake (g kg^{-1} BM)	estimated protein requirements: in diet (% of ME)
Callithrix jacchus	1	adult	0.41	soybean[b]	N-balance	4.6	7.4
Saimiri sciureus	2	2–3 week infant	0.15	casin	growth	18	15
	2	2–3 month infant	0.3	casein	growth	7.3	7.1
	2	9 month juvenile	0.5	casein	growth	4.3	5.8
	3	subadult	0.7	soybean[b]	growth[c]	8.0	12.5
Cebus albifrons	4	5–6 week infant	0.4	lactalbumin	growth	5.2	7.0
	4	3 month infant	0.6	lactalbumin	growth	4.2	6.4
	4	7 month infant	1.0	lactalbumin	growth	3.3	5.2
	5	adult	2.8	lactalbumin	weight	1.8	7.5
	4	5–6 week infant	0.4	soybean	growth	10.4	14
	4	3 month infant	0.6	soybean	growth	6.8	11
	4	7 month infant	1.0	soybean	growth	6.1	9.7
Macaca mulatta	6	1–7 month infant	1.1	milk protein	growth	3.4	5.5
Homo sapiens	7	1–6 month infant	6	milk	growth	2.2	8.0
	7	6–12 month infant	9	mixed	factorial[d]	1.6	6.6
	7	1–3 year child	13	mixed	factorial	1.2	4.9
	7	7–10 year child	28	mixed	factorial	1.0	5.6
	7	young woman	58	mixed	factorial	0.8	8.4
	7	pregnant woman	58	mixed	factorial	1.0	9.6
	7	lactating woman (1st 6 months)	58	mixed	factorial	1.1	9.6

[a] Sources of data as follows: 1, Flurer *et al.* (1988); 2, Ausman *et al.* (1979); 3, De La Inglesia *et al.* (1967); 4, Samonds & Hegsted (1973); 5, Ausman & Hegsted (1980); 6, Kerr *et al.* (1970); 7, National Research Council (1989).
[b] Methionine added to diet.
[c] Serum chemistry and hepatic histology also examined.
[d] Factorial method based on various types of published data, including N-balance.

or energy deficits, both of which increase degradation of tissue protein and hence exacerbate nitrogen losses (Torun & Viteri 1988). PEM is embedded in a web of poverty that restricts access to food, sanitation, health care, agricultural improvements and education (Oftedal & Levinson 1977), and is as much a social disease as a nutritional one.

The verdict from primate foraging studies is equally mixed. Some studies have indicated a significant correlation between food choice and protein concentration, but others have not. For example, in a study of black colobus (*Colobus satanus*) McKey *et al.* (1981) found that selection among available seeds was significantly correlated with protein concentration, but selection among mature leaves was not. Choice of leaves appears to be related to protein concentration in mantled howler monkeys (Milton 1979; Glander 1981), but not in south Indian leaf-monkeys (*Presbytis johnii*; Oates *et al.* 1980). However, virtually all investigators agree that many factors underlie food choice, and that negative factors such as indigestible fibre, digestion-inhibiting phenolic compounds and potentially toxic secondary compounds may be as important as nutrient levels.

Ultimately the importance of high-protein foods depends upon levels of protein intake relative to requirements. Protein supplies both essential amino acids that primates are unable to synthesize (or, in the case of histidine, cannot synthesize at an adequate rate during infancy), and non-essential amino acids that are readily synthesized. The latter may be important as a source of organic nitrogen, but are not individually required. The foregut fermention system of colobine monkeys presumably entails microbial metabolism and synthesis of essential amino acids, so these primates may resemble ruminants in being relatively independent of the amino acid composition of protein (Van Soest 1982).

In human studies, protein requirements are usually established in relation to reference proteins such as egg and milk proteins that contain adequate levels of essential amino acids and are highly digestible (National Research Council 1989). Because nutritional research on primates has focused on them as models for human nutrition, emphasis has been placed on requirements for reference protein rather than the proteins consumed by primates in the wild. As discussed below, the two may differ considerably.

Several methods have been utilized in estimating the protein requirements of primates (table 5). Most commonly, a high-quality milk protein such as casein or lactalbumin is fed at graded levels to infant or juvenile monkeys, and growth responses are monitored. This method will only be accurate if other factors that influence growth (especially energy intake) are controlled. Another method involves measurement of nitrogen balance (the difference between nitrogen uptake from the diet and nitrogen excreted from the body), but this method has come under criticism because of errors that may generate biased results (Hegsted 1976). Primates tend to be particularly difficult to use in balance trials due to their propensity to scatter both food and excreta, leading to measurement errors that may have a major effect on nitrogen balance estimates. Some investigators have examined the response of serum proteins, other serum constituents, or specific tissues to changes in protein intake. In all cases, protein requirements may be overestimated if ingested diets do not provide sufficient energy, as animals typically catabolize protein as a source of energy when energy intakes are low.

Unfortunately some reviews of the protein requirements of primates (see Kerr 1972; National Research Council 1978) have accepted the conclusions of investigators without sufficient appraisal of experimental design, statistical methods or speculative statements. For example, the study of Robbins & Gavan (1966) is cited to show satisfactory nitrogen balance in adult rhesus even though individual values for nitrogen retention ranged from −31% to +28%. This was a very short study (4 days) with limited pre-adaptation of the rhesus to experimental conditions. Similarly, the statement of Hodson *et al.* (1967) that the lowest protein level that they fed to growing chimpanzees may have been marginal was accepted even though there were no treatment effects on growth rate, nitrogen balance or serum proteins. Many of the early studies on protein requirements of primates are difficult to evaluate as they include little or no statistical analysis.

The protein requirements of primates have typically been expressed in one of two ways, either as a daily intake in relation to body mass (grams per kilogram), or as a percentage of dietary metabolizable energy (table 5). Comparisons among species are complicated by differences in body size, age class and type of protein used in requirement studies. Within a species, the requirements of growing infants and juveniles appear to decline with an increase in age and body mass, whether protein requirements are expressed relative to body mass or as a percentage of dietary energy (table 5). Unfortunately there are relatively few data on protein requirements of adult primates. In humans, the estimated protein requirement of an adult female is lower (0.8 g kg^{-1}) than that of a child (1.0–1.2 g kg^{-1}) when expressed relative to body mass. However recommended energy intakes decrease even more (from about 90–100 kcal kg^{-1} in a young child to 38 kcal kg^{-1} in a young woman; National Research Council 1989) such that protein requirements expressed as a percentage of energy actually increase (from 4.9 to 8.4%; table 5). A similar phenomenon has been observed in captive *Cebus* fed a reference protein (lactalbumin): relative to body mass the adult requirement (1.8 g kg^{-1}) is lower than that of the juvenile (3.3 g kg^{-1}), but as a percentage of energy the adult requirement is higher (7.5% versus 5.2%).

It is apparent that young infants have an especially high requirement for protein, but at this time they consume predominantly mother's milk. The proteins in primate milks are presumably equivalent or superior to the reference proteins (casein and lactalbumin from cow's milk) that have been tested, and typically supply 7–22% of the energy (Oftedal 1984, O. T. Oftedal, unpublished data). It is not known whether the higher protein concentration (20–22% of energy) observed in the milks of some primate species is indicative of particularly high protein requirements of infants, but it seems likely given the cost to the mother of producing milk that is high in protein.

Unfortunately, virtually no research has been conducted on protein requirements during pregnancy and lactation in primates, other than studies of the effects of pronounced protein deficiency on females and their infants (see, for example, Riopelle *et al.* (1975); Kohrs *et al.* (1980)). In humans it is estimated that protein requirements increase by more than one third (from 0.8 to 1.1 g kg^{-1} body mass; table 5) during early lactation, and an even greater increase probably accompanies the onset of lactation in species that produce milks higher in protein concentration. Human milk has the lowest protein concentration (about 7% of energy) of any primate milk that has been studied. In general, it appears that primates produce small daily amounts of a relatively dilute milk (Oftedal 1984). Thus the protein and energy demands of lactation are probably low for primates by comparison to the demands experienced by many other mammals.

Proteins other than reference proteins are usually required at considerably higher levels to compensate for lower digestibility and shortfalls in essential amino acids. For example, the apparent digestibility of plant proteins consumed by humans may be as little as 78% (beans) or as much as 96% (refined wheat), but for most plant products is about 85–88% (National Research Council 1989). In cebus monkeys (*Cebus albifrons*) the true digestibility of lactalbumin is about 100%, but that of soy protein concentrate is only 83% (Ausman *et al.* 1986). By complexing with proteins, tannins may reduce digestibility of plant proteins even further (Swain 1979). Although tannins are particularly abundant in leaves, they are also found in some fruits (including seeds), stems and flowers eaten by primates (see, for example, McKey *et al.* (1981); Calvert (1985); Barton *et al.* (1992)).

In any protein the essential amino acid that is in least supply relative to the levels in reference protein determines the amino acid score of that protein. A score of 50% indicates that the amount of protein that must be consumed to meet protein requirements is twice (1/0.50) that of the reference protein. The protein in cereal grain can have an amino acid score of less than 50% (owing to low lysine concentration) whereas the amino acid score of protein from legume seeds may be only 70% (due to low concentration of the sulphur-containing amino acids, cysteine and methionine) (Munro & Crim 1988). Studies of young cebus monkeys have shown that soy protein concentrate has a potency of only 53–69% (compared to lactalbumin) in supporting growth and nitrogen retention (Ausman *et al.* 1986). Thus estimated protein requirements for growing cebus are considerably higher when they are fed soy protein (without amino acid supplementation) in place of lactalbumin (table 5). However, even when soy protein is supplemented with methionine (a sulphur-containing amino acid), it is usually less potent than reference protein, presumably due to lower digestibility.

Relatively little is known about the essential amino acid patterns in foods consumed by wild primates. Glander (1981) reported higher levels of most essential amino acids in leaves eaten by mantled howlers than in leaves that were not eaten, but this difference disappears once the amino acid data are expressed as a percentage of protein. As a first estimate of amino acid adequacy, the amino acid composition of mature and young leaves eaten by howlers (Glander 1981) may be compared to the estimated amino acid requirements of 2-year-old-children (National Research Council 1989). Expressed as a percentage of protein, the average levels of histidine, isoleucine, leucine, lysine, phenylalanine + tyrosine, threonine, and valine all exceed the NRC figures for requirements. Methionine levels (1.8–2.0% of protein) are somewhat lower than total sulphur-containing amino acid requirements (2.5%), but cysteine was not measured. If the cysteine concentration of the protein in these leaves is similar to that of most leaf proteins (i.e. about 0.7–1.4%; Lyttleton 1973), the level of total sulphur-containing amino acids should not be limiting. Although tryptophan analysis was also omitted, leaf proteins usually contain 1.6–2.1% tryptophan (Lyttleton 1973), well above the NRC suggested requirement of 1.1%. Because leaves typically contain abundant levels of essential amino acids, the adequacy of leaf proteins for animal feeding is impaired more by the presence of tannins, growth depressants and toxic compounds than by amino acid imbalance (Allison 1973). Inappropriate amino acid levels are more likely to occur in storage organs such as tubers and seeds (Van Soest 1982).

What can we conclude about the protein requirements of primates in captivity and the wild? Once the period of milk dependency is past, growing and non-reproductive adult primates appear to require 5–8% of metabolizable energy as protein, if reference protein is fed (table 5). For manufactured feeds based on plant proteins but supplemented with limiting amino acids (e.g. methionine or lysine), this value should be increased to 6–9% of ME (7–10% of dry matter (DM)) on the assumption that protein digestibility is about 85%. A pregnant or lactating female may require at least 10% of ME as reference protein, equivalent to a level of not less than 12.5% protein (DM basis). The NRC recommendation is 15% protein (16% DM basis) for all stages of life (table 3). The suggestion that New World monkeys may require up to 25% protein (National Research Council 1978) is not supported by available data (table 5).

Estimation of the protein requirement of primates that feed on natural foods is complicated by uncertainties about protein digestibility, amino acid patterns and metabolizable energy concentration. Assuming a protein digestibility of 85%, an amino acid score of 100%, and a metabolizable energy concentration of only 3 kcal ME per gram DM (due to high fibre concentration and low fibre digestibility), a primate eating leaves would require 4–7% protein (DM basis) for growth and maintenance, and at least 8% (DM basis) for reproduction. However, with the same energy concentration and only 50% protein digestibility (due to severe effects of tannins), the estimated requirements would be 7–11% (DM basis) for growth and maintenance and at least 14% (DM basis) for reproduction. Given that the leaves eaten by primates usually average 12–16% protein (see, for example, Glander 1981; Calvert 1985; table 1), it is unlikely that protein deficiency will be a problem except perhaps for lactating females consuming leaves of high tannin content. Foods with unbalanced amino acid patterns may present a different picture. For example, if high-fat seeds have 85% protein digestibility, 60% amino acid score and 4.5 kcal per gram DM, the estimated requirements would be 10–16% and at least 20% protein (DMB) for growth and maintenance, and reproduction, respectively.

However, caution is required in interpreting analytical data on protein. Crude protein concentration is calculated, by convention, as total nitrogen (TN) times a factor (6.25) that assumes that protein contains 16% nitrogen. However, plants contain non-protein nitrogenous compounds, including alkaloids and non-protein amino acids, some of which cannot be metabolized by animals (Munro & Crim 1988). Some plant proteins have a higher concentration of nitrogen than 16%. The correct factor for calculating protein from nitrogen concentration is about 5.8 for many grains, 5.3–5.7 for some leguminous seeds (e.g. soybeans and peanuts), and 5.2–5.3 for other seeds and nuts (Watt & Merrill 1963). The appropriate factor for tropical leves may be even lower (*ca.* 4.0–5.0; Milton & Dintzis 1981). Thus crude protein (TN × 6.25) may substantially overestimate true protein.

7. CONCLUSION

Nutrients that are consumed at marginal or inadequate levels with respect to requirements may limit animal performance (and, ultimately, evolutionary fitness). Theoretically, animals should evolve feeding behaviours that enhance intakes of limiting nutrients. However, there may be little or no advantage to enhancing intake of a nutrient that is already abundant relative to requirements.

Despite the large number of studies that have been undertaken to relate food selection to food chemistry, very little information is available about the nutrient intakes of primates consuming natural diets or how these intakes relate to nutrient requirements. Further research is needed on feeding rates, digestibilities of foods, energy utilization and nutrient requirements, especially in relation to reproduction.

Because primates typically grow slowly and have low daily milk yields, age and reproductive effort probably have less effect on nutrient requirements than in most mammals. This reproductive strategy may be important in allowing primates to use foods of only moderate nutrient density on a year-round basis.

Research on the nutrition of red howler monkeys was funded by the Smithsonian International Environmental Studies Program. I thank M. Klein, M. Power, A. Rosenberg and A. Whiten for helpful comments on the manuscript.

REFERENCES

Allen, M. E. 1989 Nutritional aspects of insectivory. Ph.D. thesis, Michigan State University.

Allison, R. M. 1973 Leaf protein as an animal and human foodstuff. In *Chemistry and biochemistry of herbage* (ed. G. W. Butler & R. W. Bailey), pp. 61–79. New York: Academic Press.

Altmann, S. A., Post, D. G. & Klein, D. F. 1987 Nutrients and toxins of plants in Amboseli, Kenya. *Afr. J. Ecol.* **25**, 279–293.

Ausman, L. M. & Hegsted, D. M. 1980 Protein requirements of adult cebus monkeys (*Cebus albifrons*). *Am. J. clin. Nutr.* **33**, 2551–2558.

Ausman, L. M., Gallina, D. L., Sammonds, K. W. & Hegsted, D. M. 1979 Assessment of the efficiency of protein utilization in young squirrel and macaque monkeys. *Am. J. clin. Nutr.* **32**, 1813–1823.

Ausman, L. M., Gallina, D. L., Hayes, K. C. & Hegsted, D. M. 1986 Comparative assessment of soy and milk protein quality in infant Cebus monkeys. *Am. J. clin. Nutr.* **43**, 112–127.

Barnes, R. F. 1973 Laboratory methods of evaluating feeding value of herbage. In *Chemistry and biochemistry of herbage* (ed. G. W. Butler & R. W. Bailey), pp. 179–214. New York: Academic Press.

Barton, R. A., Whiten, A., Byrne, R. W. & English, M. 1992 Chemical composition of baboon food plants: implications for the interpretation of intra- and interspecific differences in diet. *Folio primatol.* (In the press.)

Bauchop, T. & Martucci, R. W. 1968 Ruminant-like digestion of the langur monkey. *Science, Wash.* **161**, 698–700.

Calvert, J. J. 1985 Food selection by western gorillas (*G. G. gorilla*) in relation to food chemistry. *Oecologia, Berl.* **65**, 236–246.

Cheeke, P. R. & Shull, L. R. 1985 *Natural toxicants in feeds and poisonous plants*. Westport, Connecticut: AVI Publishing Co.

Crissey, S. D., Edwards, M. S., Oftedal, O. T., Currier, J. A. & Rudran, R. 1991 The role of fiber in natural and manufactured diets fed to red howler monkeys (*Alouatta seniculus*). In *Proceedings of the Eighth Dr Scholl Conference on the Nutrition of Captive Wild Animals* (ed. T. P. Meehan, S. D. Thompson & M. E. Allen), pp. 135–147. Chicago, Illinois: Lincoln Park Zoological Society.

De La Inglesia, F. A., Porta, E. A. & Hartroft, W. S. 1967 Effects of dietary protein levels on the *Saimiri sciureus*. *Expl molec. Pathol.* **32**, 1813–1823,

FAO 1965 *Protein requirements* (Report of a joint FAO/WHO expert group). Rome: Food and Agriculture Organization.

Flurer, C. I., Krommer, G. & Zucker, H. 1988 Endogenous N-excretion and minimal protein requirement for maintenance of the common marmoset (*Callithrix jacchus*). *Lab. Anim. Sci.* **38**, 183–186.

Glander, K. E. 1981 Feeding patterns in mantled howling monkeys. In *Foraging behavior: ecological, ethological, and psychological approaches* (ed. A. C. Kamil & T. D. Sargent), pp. 231–257. New York: Garland Press.

Glander, K. E. 1982 The impact of plant secondary compounds on primate feeding behavior. *Yb. Phys. Anthro.* **25**, 1–18.

Harding, R. S. O. 1981 An order of omnivores: Nonhuman primates in the wild. In *Omnivorous primates. Gathering and hunting in human evolution* (ed. R. S. O. Harding & G. Teleki), pp. 191–214. New York: Columbia University Press.

Hegsted, D. M. 1976 Balance studies. *J. Nutr.* **106**, 307–311.

Hladik, C. M. 1977 A comparative study of the feeding strategies of two sympatric species of leaf monkeys: *Presbytis senex* and *Presbytis entellus*. In *Primate ecology: studies of feeding and ranging behaviour in lemurs, monkeys and apes* (ed. T. H. Clutton-Brock), pp. 323–353. New York: Academic Press.

Hodson, H. H., Mesa, V. L. & Van Riper, D. C. 1967 Protein requirement of the young, growing chimpanzee. *Lab. Anim. Care* **17**, 551–562.

Janson, C. H. 1988 Intra-specific food competition and primate social structure: a synthesis. *Behaviour* **105**, 1–17.

Kerr, G. R. 1972 Nutritional requirements of subhuman primates. *Physiol. Rev.* **52**, 415–467.

Kerr, G. R., Allen, J. R., Scheffler, G. & Waisman, H. A. 1970 Malnutrition studies in the rhesus monkey. I. Effect on physical growth. *Am. J. clin. Nutr.* **23**, 739–748.

Kohrs, M. B., Kerr, G. R. & Harper, A. E. 1980 Effects of a low protein diet during pregnancy of the rhesus monkey. III. Growth of infants. *Am. J. clin. Nutr.* **33**, 625–630.

Lyttleton, J. W. 1973 Proteins and nucleic acids. In *Chemistry and biochemistry of herbage* (ed. G. W. Butler & R. W. Bailey), pp. 63–103. New York: Academic Press.

Martin, R. D., Chivers, D. J., MacLarnon, A. M. & Hladik, C. M. 1985 Gastrointestinal allometry in primates and other mammals. In *Size and scaling in primate biology* (ed. W. L. Jungers), pp. 61–89. New York: Plenum Press.

McKey, D. B., Gartlan, J. S., Waterman, P. G. & Choo, G. M. 1981 Food selection by black colobus monkeys (*Colobus satanus*) in relation to plant chemistry. *Biol. J. Linn. Soc.* **16**, 115–146.

Milton, K. 1979 Factors influencing leaf choice by howler monkeys: a test of some hypotheses of food selection by generalist herbivores. *Am. Nat.* **114**, 362–378.

Milton, K. 1980 *The foraging strategy of howler monkeys*. New York: Columbia University Press.

Milton, K. 1981 Food choice and digestive strategies of two sympatric primate species. *Am. Nat.* **117**, 496–505.

Milton, K. 1984 The role of food-processing factors in primate food choice. In *Adaptations for foraging in non-human primates* (ed. P. S. Rodman & J. G. H. Cant), pp. 249–279. New York: Columbia University Press.

Milton, K. & Demment, M. W. 1988 Digestion and passage kinetics of chimpanzees fed high and low fiber diets and comparison with human data. *J. Nutr.* **118**, 1082–1088.

Milton, K. & Dintzis, F. R. 1981 Nitrogen-to-protein conversion factors for tropical plant samples. *Biotropica* **13**, 177–181.

Milton, K. & McBee, R. H. 1983 Rates of fermentative digestion in the howler monkey, *Alouatta palliata* (Primates, Ceboidea). *Comp. Biochem. Physiol.* A **74**, 29–31.

Milton, K., Van Soest, P. J. & Robertson, J. 1980 The digestive efficiencies of wild howler monkeys. *Physiol. Zool.* **53**, 402–409.

Morris, E. R. 1986 Phytate and dietary mineral bioavailability. In *Phytic acid: chemistry and applications* (ed. E. Graf), pp. 57–76. Minneapolis, Minnesota: Pilatus Press.

Munro, H. N. & Crim, M. C. 1988 The proteins and amino acids. In *Modern nutrition in health and disease* (ed. M. E. Shils & V. R. Young), pp. 1–37. Philadelphia, Pennsylvannia: Lea & Febiger.

National Research Council 1978 *Nutrient requirements of nonhuman primates*. Washington, DC: National Academy of Sciences.

National Research Council 1980 *Mineral tolerance of domestic animals*. Washington: National Academy of Sciences.
National Research Council 1989 *Recommended Dietary Allowances*, 10th edn. Washington, D.C.: National Academy Press.
Oates, J. F., Waterman, P. G. & Choo, G. M. 1980 Food selection by the South Indian leaf-monkey, *Presbytis johnii*, in relation to leaf chemistry. *Oecologia, Berl.* **45**, 45–56.
Oftedal, O. T. 1984 Milk composition, milk yield and energy output at peak lactation: a comparative review. *Symp. zool. Soc. Lond.* **51**, 33–85.
Oftedal, O. T. & Levinson, F. J. 1977 Equity and income effects of nutrition and health care. In *Income distribution and growth in less-developed countries* (ed. C. R. Frank & R. C. Webb), pp. 381–433. Washington, DC: Brookings Institution.
Oftedal, O. T., Jakubasz, M. & Whetter, P. 1982 Food intake and diet digestibility by captive black and white colobus (*Colobus guereza*) at the National Zoological Park. In *Proceedings, 1982 Annual Meeting, American Association of Zoo Veterinarians*. New Orleans, Louisianna.
Power, M. L. 1991 Digestive function, energy intake and the response to dietary gum in captive callitrichids. Ph.D. thesis, University of California at Berkeley.
Riopelle, A. J., Hill, C. W., Li, S., Wolf, R. H., Seibold, H. R. & Smith, J. L. 1975 Protein deficiency in primates. IV. Pregnant rhesus monkey. *Am. J. Clin. Nutr.* **28**, 20–28.
Robbins, C. & Gavan, J. A. 1966 Utilization of energy and protein of a commercial diet by rhesus monkeys (*Macaca mulatta*). *Lab. Anim. Care* **16**, 286–291.
Samonds, K. W. & Hegsted, D. M. 1973 Protein requirements of young cebus monkeys (*Cebus albifrons* and *apella*). *Am. J. clin. Nutr.* **26**, 30–40.
Swain, T. 1979 Tannins and lignins. In *Herbivores: their interaction with secondary plant metabolites* (ed. G. A. Rosenthal & D. H. Janzen), pp. 657–682. New York: Academic Press.
Torun, B. & Viteri, F. E. 1988 Protein-energy malnutrition. In *Modern nutrition in health and disease* (ed. M. E. Shils & V. R. Young), pp. 746–773. Philadelphia, Pennsylvania: Lea and Febiger.
Van Soest, P. J. 1982 *Nutritional ecology of the ruminant*. Corvallis: O & B Books, Inc.
Watkins, B. E., Ullrey, D. E. & Whetter, P. A. 1985 Digestibility of a high-fiber biscuit-based diet by black and white colobus (*Colobus guereza*). *Am. J. Primatol.* **9**, 137–144.
Watt, B. K. & Merrill, A. L. 1963 *Composition of foods* (Agriculture Handbook No. 8). Washington, D.C.: U.S. Department of Agriculture.

Discussion

A. Whiten (*Scottish Primate Research Group, University of St Andrews, U.K.*). Dr Oftedal lists nutritional factors known to affect the value of a food. Should not the occurrence of secondary compounds in plant foods, some of which influence digestibility directly, be added to that list?

O. T. Oftedal. Secondary compounds in plants may produce toxicity or reduce digestibility of fibre, protein or other nutrients in farm and laboratory animals, but little is known about their effects on primates. Unfortunately, adverse effects are hard to predict. Most screening tests for secondary compounds in foods are qualitative rather than quantitative, but toxic effects are dose-dependent. Moreover, most tests are for classes of compounds (e.g. alkaloids, saponins, tannins, cyanogenetic glycosides) even though toxicity or digestive effects may vary greatly between one compound and the next. The only way to ascertain the effects of secondary compounds is to conduct controlled toxicological and digestibility studies using the specific compounds and primate species of interest, but to my knowledge this has not been done.

D. A. T. Southgate (*AFRC Institute of Food Research, Norwich, U.K.*). I would like to reinforce the comments of Professor Altmann [comments not supplied] about the extent to which the complex carbohydrates in plant cell walls (fibre) are degraded in the large bowel of humans. It is important to distinguish between the different sources and composition of these complex carbohydrates, as these will influence the energy that the animal obtains. The acid lignin values may include other substances, for example cutin, and may exaggerate the extent to which the materials resist degradation.

O. T. Oftedal. The amount of energy that can be obtained by a primate from the complex carbohydrates in plant cell walls will depend both on the types of carbohydrates and the species of primate. Humans are probably not very representative of most primates because of their large body size. In mammals, large size usually correlates with slower passage rates of digesta with consequent increased opportunity for fermentation. However, some primates have evolved specialized fore- or hind-gut fermentation systems, and are undoubtedly more efficient at fibre fermentation than humans.

P. Van Soest (*324 Morrison Hall, Cornell University, New York, U.S.A.*). I comment on the high lignin content of leaves. It is probably not true lignin or cutin, but rather tannin, which often confounds crude lignin analysis. The effects of tannin upon reduction of fibre fermentation is low as compared with lignin, but much greater for the case of protein digestion. Sequential analysis can help resolve this contamination of crude lignin. Extractions in the sequential order: neutral detergent, acid detergent followed by lignin determination (yielding a maximum value) can be compared with the sequence: acid detergent, neutral detergent and lignin isolation. The difference between lignin values by the first sequence and second sequences gives some estimation of insoluble condensed tannins.

O. T. Oftedal. I agree that further investigation of the various fractions obtained with the detergent fibre system is needed, and believe the comparison Dr Van Soest suggests would be valuable.

P. Van Soest. Polysaccharides like pectins, fructans and galactans, which are soluble and highly fermentable, but not digestible by mamalian digestive enzymes, yield bacteria and volatile fatty acids from colonic fermentation and about 3 cal g^{-1} of metabolizable energy.

O. T. Oftedal. Such polysaccharides are of particular importance to primates that specialize on plant exudates, such as marmosets and some galagos.

The significance of fibrous foods for Kibale Forest chimpanzees

R. W. WRANGHAM, N. L. CONKLIN, C. A. CHAPMAN AND K. D. HUNT

Department of Anthropology, Peabody Museum, Harvard University, Cambridge, Massachusetts 02138, U.S.A.

SUMMARY

Four categories of plant food dominated the diet of chimpanzees in Kibale Forest, Uganda: non-fig tree fruits, fig tree fruits, herbaceous piths and terrestrial leaves. Fruit abundance varied unpredictably, more among non-figs than figs. Pith intake was correlated negatively with fruit abundance and positively with rainfall, whereas leaf intake was not influenced by fruit abundance. Piths typically have low sugar and protein levels. Compared with fruits and leaves they are consistently high in hemicellulose and cellulose, which are insoluble fibres partly digestible by chimpanzees. Herbaceous piths appear to be a vital resource for African forest apes, offering an alternative energy supply when fruits are scarce.

1. INTRODUCTION

Chimpanzees (*Pan troglodytes*) are primarily frugivores (Ghiglieri 1984; Hladik 1977; Isabirye–Basuta 1990; McGrew *et al.* 1988; Nishida & Uehara 1983; Wrangham 1977). When fruit is rare, chimpanzees, like other frugivores, must migrate to more productive areas (Nishida 1979), reduce energy expenditure (Wrangham 1977), or broaden their diet. Here the nutritional constituents of piths and their pattern of utilization is examined to test the hypothesis that chimpanzees rely on fibrous piths when fruit is scarce.

Like other African apes, chimpanzees eat piths, primarily from herbaceous stems in the ground layer. The number of species eaten varies from two (Mount Assirik, Senegal (McGrew *et al.* 1988)) to 28 (Mahale Mountains, Tanzania (Nishida & Uehara 1983)). Chimpanzees select stems that are typically more than 2 cm thick. They use their teeth to break the tough outer peel and extract the softer central pith. Unlike most other primates and ungulates, African apes have teeth that appear effective at shearing such stems, because of their large size, thin enamel, and long cutting edges (Kay 1981). In addition, the relatively efficient digestion of high-fibre foods in captive chimpanzees, presumably due to their large hindgut and total gut volumes (Milton & Demment 1988), suggests that chimpanzees may have special adaptations for eating piths.

Together with leaves eaten in the ground layer, piths have been reported to provide critical fallback foods in three studies of bonobos (pygmy chimpanzees, *Pan paniscus*) and one of frugivorous gorillas (*Gorilla gorilla*) when fruits were thought to be scarce (Badrian & Malenky 1984; Kano 1983; Kano & Mulavwa 1984; Rogers *et al.* 1988). In habitats with little fruit, terrestrial piths and leaves (TPL) provide the principal components of gorilla diet (Calvert 1985; Goodall 1977; Watts 1984).

In contrast to the importance of terrestrial piths and leaves in sustaining bonobos and gorillas, their significance for wild chimpanzees during periods of fruit shortage is little known. Instead of relying on piths and leaves, present indications are that fruitless chimpanzees resort to a variety of low-quality items (e.g. bark (Nishida 1976)). In the only study relating dietary changes to phenological measures of fruit abundance, Isabirye–Basuta (1990) found that when preferred fruits were scarce, Kibale Forest chimpanzees increased their diet diversity without emphasizing any particular class of food such as leaves. Similarly, comparisons between geographical areas suggest that poor food conditions favour a generalized increase in diet diversity involving stems, barks, underground storage organs and insects (see, for example, McGrew *et al.* 1988; Nishida 1989; Suzuki 1969).

In this paper we investigate the use of terrestrial fibrous foods by unprovisioned chimpanzees in the Kibale Forest. These foods have previously been designated as THV (terrestrial herbaceous vegetation) (Rogers & Williamson 1987; Wrangham 1986). Here we call them TPL (terrestrial piths and leaves) because of the occasional importance of woody species. We ask how TPL intake is related to both fruit abundance and rainfall, and whether TPL, including herbaceous pith, herbaceous leaves and shrub leaves, should be regarded as substitutes for fruits, because of sugar components, as suggested by Badrian *et al.* (1981) and Kano (1983) or as complementary because of their protein value, as proposed by Hladik (1977) and Malenky (1990).

2. STUDY SITE AND METHODS

We observed chimpanzees in the Kibale Forest Reserve, western Uganda (0° 13′–0° 41′ N, 30° 19′–30° 32′ E), from December 1987 to March 1991 (Wrangham *et al.* 1993). The Kanyawara community was the principal study group, comprising about 50 individuals in more than 20 km^2, (Isabirye-Basuta 1990). Kanyawara has undulating ground and an elevation of $\approx$ 1500 m. The vegetation is a mosaic of mid-altitude moist forest, secondary forest, tall grassland, swamps and softwood plantations. Supporting data came from the Ngogo community (Ghiglieri 1984) and sympatric Ngogo baboons (*Papio anubis*). Ngogo is 10–15 km S.E. of Kanyawara in similar habitat $\approx$ 150 m lower, without logged forest or softwood plantations (Butynski 1990). Feeding records were collected by direct observation, by inspection of feeding remains, and by dung analysis. Observations were made by the authors and field assistants (Wrangham *et al.* 1992). Direct observations reported here were by R. W. W.

Monthly fruit abundance was estimated in Kanyawara by recording the phenological state of 227 trees on a 12 km transect every two weeks from December 1987 to March 1990. Species observed were those seen to be important chimpanzee foods from 1983 to 1985 (G. Isabirye-Basuta, personal communication). Non-fig fruit trees were 20 *Celtis africana*, 20 *C. durandii*, 20 *Tabernaemontana* spp. (*T.* (*Conopharyngia*) *holstii* and *T.* (*Gabunia*) *odoratissima*), 9 *Cordia abyssinica*, 7 *Monodora myristica*, 11 *Mimusops bagshawei*, 12 *Pseudospondias microcarpa*, 20 *Teclea nobilis* and 20 *Uvariopsis congensis*. Fig trees were 15 *F. asperifolia* Miq., 5 *F. conraui* Warb., 20 *F. exasperata* Vahl, 9 *F. natalensis* Hochst., 1 *F. sansibarica* Warb., subsp. *macrosperma* (Mildbr. & Burret) C. C. Berg (type 1), 14 *F. s. macrosperma* (type 2) and 4 *F. saussureana* DC. Fig nomenclature follows Berg and Hijman (1989). To obtain fruit availability indices (FA) for each species, we multiplied the percentage of trees with ripe fruit by the density of trees in the study area (found from a stratified sample of 2300 trees) and by the mean basal area of each reproductive stem (also throughout the study area). FA indices were summed across species, and numbers were adjusted to make the largest FA index equal to 100 (i.e. FA (all) in November 1989).

Three fruit abundance indices were calculated by summing across different groups of tree species, i.e. FA (non-fig), FA (fig) and FA (all). Because *F. exasperata* fruits were sometimes not eaten by chimpanzees even when abundant (Isabirye-Basuta 1990), we also calculated FA (all except *F. exasperata*). This measure behaved almost identically to FA (all).

Food samples (only parts eaten by chimpanzees) were air-dried in the field. From this, room-temperature dry matter was calculated. Further analysis was done in the nutritional biochemistry laboratory in the Anthropology Department at Harvard University.

Standard chemical analyses were done to estimate nutritional value. Crude protein was determined using Kjeldahl procedure for total nitrogen, and multiplied by 6.25 (Pierce & Haenisch 1947). The digestion mix contained Na_2SO_4 and $CuSO_4$. The distillate was collected in 4% (by volume) boric acid and titrated with 0.1 N HCl. The detergent system of fibre analysis (Goering & van Soest 1970), as modified by Robertson & van Soest (1980), was used to determine the neutral-detergent fibre, hemicellulose, cellulose and lignin fractions. The lignin determination was done with 72% suphuric acid. Total lipid content was measured using petroleum ether extraction for four days at room temperature (modified from AOAC 1984). Water-soluble carbohydrates were estimated using phenol-sulphuric acid colorimetric assay (Dubois *et al.* 1956), as modified by Strickland & Parsons (1972). Condensed tannin content was measured using the proanathocyanidin test of Bate-Smith (1975) as presented by Mole & Waterman (1987). Dry matter was determined by drying a subsample at 100 °C for 8 h and hot weighing. Total ash was measured by ashing the above subsample at 520 °C for 8 h and then hot weighing at 100 °C (van Soest & Robertson 1991).

Statistical tests are two-tailed except where stated.

3. COMPOSITION OF CHIMPANZEE DIET

The plant diet of Kanyawara chimpanzees fell into three principal categories: tree fruits other than figs (21 species identified to date), fig-tree fruits (10 species), and TPL (14 pith species, 28 leaf species). Additional categories were fruits from shrubs, vines or herbs (15 species), leaves from mature trees (1 species), seeds (1), flowers (1) and bark (1). This list is certainly incomplete as we have found seeds of at least 20 unidentified fruit species in the dung.

We rely principally on dung analysis for comparisons of food intake over time or between populations to overcome sampling biases present in direct observation of food intake: chimpanzees were more easily discovered when eating tree fruits, because they returned repeatedly to productive trees and frequently gave loud calls. By contrast, we often lost contact with chimpanzees when they began to eat TPL, because when doing so they selected their foraging areas unpredictably, tended to be silent, and were rarely visible (the vegetation is often dense and tall).

(*a*) *Non-fig tree fruits*

Fruit eating occupied the majority of feeding time, and during any month the fruit diet was dominated by only one or a few species. During 14 days in 1991, each with at least 5 h of focal observation per day, 59.7% of time was spent feeding: chimpanzees ate arboreal fruits for 71.7% of feeding time ($n = 100.5$ h; $n = 4$ days in February, 4 days in March and 6 days in April). Fruits in this sample came from six species of tree, two of which predominated (*M. bagshawei* 66.0% of fruit-eating time, *Ficus natalensis* 23.2%). These figures appear typical of Kibale chimpanzees (Ghiglieri 1984; Isabirye-Basuta 1990).

Non-fig tree fruits of preferred species were eaten frequently when available. The most important fruit species at Kanyawara was *M. bagshawei* S. Moore, which was eaten in 14 out of 35 months. Its seeds occurred in an average of 22.1% of dungs in all months (range 0–94.1%). Phenological data for *M. bagshawei* fruits show that they were selected as a function of their availability (figure 1). In Kanyawara only five other species produced fruits whose residues were found in at least 50% of dungs in any month (*Cordia abyssinica*, *Monodora myristica*, *Pseudospondias microcarpa*, *Taberna-*

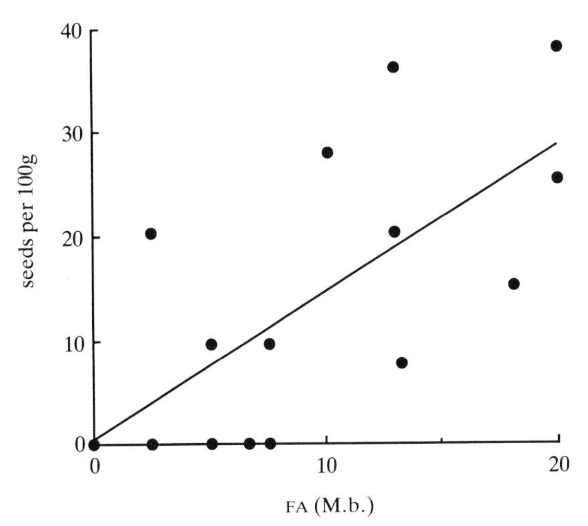

Figure 1. Frequency of eating *Mimusops bagshawei* S. Moore fruits in relation to their availability. 'FA (M.b.)' = fruit availability index of *M. bagshawei* ripe fruits. Each point is one month. 'Seeds per 100 g' = mean number of *M. bagshawei* seeds found per 100 g of chimpanzee dung (wet mass). (The mean mass of a *M. bagshawei* seed is 0.6 g, representing 1.8 g of fruit pulp.) Data are from December 1987 to March 1990. The slope of the linear regression is shown ($r^2 = 0.62$, d.f. = 27, $p < 0.001$).

emontana holstii and *T. odoratissima* (not distinguished in the field), and *Uvariopsis congensis*). The equivalent list for Ngogo was similarly small: *Chrysophyllum albidum*, *M. bagshawei*, *M. myristica*, *U. congensis* and *Warburgia ugandensis*. Each species fruited irregularly (eaten for a mean of 3.5 months per year), and because the number of potentially important tree-fruit species was small there were times when few tree-fruits were available (e.g., in 1989 only figs were present for four months). Fruits from lianas, vines and herbs are generally taken in small amounts.

(*b*) *Fig-tree fruits*

Figs were nearly continuously available (see below) and were eaten throughout the study period (cf. Wrangham *et al.* 1993). Fig seeds (predominantly from fig trees) occurred in 93.7 % of Kanyawara dungs (n = 839) (cf. 94.9 % of Ngogo dungs, n = 416), and were present in dungs in all months (Kanyawara: 35 months, December 1987–October 1990; Ngogo: 32 months, February 1988–October 1990). The abundance of fig seeds in dung was scored on a 0–4 scale (i.e. zero, rare, few, common, abundant). Mean scores were nearly equivalent at Kanyawara (mean = 2.54, n = 839) and Ngogo (mean = 2.68, n = 416).

(*c*) TPL

During our February–April 1991 sample, focal chimpanzees spent 17.6 % of their feeding time eating TPL, including occasions when we interpreted their activity from the sounds of breaking stems. The mean duration of 27 TPL-eating sessions was 24.0 min (s.d. 46.7), with 2.8 food types eaten per session (s.d. 2.0).

Stands of herbs whose pith is eaten vary from isolated individuals to high density, monospecific 'fields'. Pith fields occur especially in gaps or at forest edges, are available year round and can cover 0.1 ha† or more. Although pith fields tend to be dominated by a single species, such as *Pennisetum purpureum* or *Acanthus pubescens*, they normally contain a variety of edible items including the piths, young leaves and fruits of several species. Thus, ten or more food items may be selected during a single feeding session.

In Kanyawara the herbs whose pith is eaten most often are *Pennisetum purpureum* (elephant grass), *Aframomum mala and Marantochloa leucantha*. Chimpanzees also eat piths from shrubs, especially *Acanthus pubescens*. Piths from at least one of these four species are probably eaten almost daily. Piths are occasionally eaten from the saplings of trees or from the midrib of palm fronds.

TPL leaves eaten by chimpanzees were invariably young, and came from herbs (8 species), shrubs (3 species), vines (3 species), or saplings (14 species). Leaves typically were eaten for less time than pith, but the leaves of *Acalypha ornata* and *Ficus urceolaris* (a shrub) were sometimes the dominant food item.

Rates of pith intake were calculated on five occasions, in bouts lasting from 1 to 23 min. The amount of food eaten was reconstructed by measuring the remains of the peel. Calculated intake rates varied from 5 to 54 g (wet mass) per minute. This is very similar to the range for fruit intake (Wrangham *et al.* 1993). There is no indication in either case that intake rates influence selectivity.

To compare the frequency of TPL feeding between Kanyawara and Ngogo, dungs were scored on a 0–4 scale for abundance of fibrous strands (FBR), and green leaf fragments (GLF). The easily identifiable, long fibrous strands, which we assume came only from eating herbaceous piths, occurred in 93.8 % of Kanyawara dungs (n = 839) and 94.9 % of Ngogo dungs (n = 416). Over 35 months (December 1987 to October 1990) the mean monthly FBR score varied at Kanyawara between 0.9 (approximately 0.5 % of dung wet mass) and 3.8 (*ca.* 10 % of dung wet mass), similar to the range at Ngogo (0.0–3.5). Rates of eating fibrous foods were thus high, and similar between Kanyawara and Ngogo.

GLFs were less common: they occurred in 28.4 % of Kanyawara chimpanzee dungs, as against 18.2 % for Ngogo chimpanzees, and had low mean scores (0.4 at both sites, *ca.* 0.2 % of dung wet mass). The maximum mean score in any month was 1.0 at Kanyawara and 0.8 at Ngogo. FBR and GLF scores were not correlated with each other either at Kanyawara (r = 0.16, n = 35, n.s.) or Ngogo (r = 0.18, n = 21, n.s.). The dung data therefore suggest that intake of pith and leaves varied in different ways.

(*d*) *Tree leaves*

Leaves are generally eaten from immature trees (for example, *Celtis africana* and *C. durandii*). Leaves of mature trees are eaten only rarely (e.g. observed once from February to April 1991), although *Ficus exasperata*

† 1 ha = 10^4 m^2.

leaves can be eaten regularly at times (Isabirye–Basuta 1990).

(*e*) *Miscellaneous plant items*

Seeds, flowers, wood and bark are eaten. They are never predominant, nor are any items eaten regularly. Seeds are mostly immature winged seeds of *Pterygota mildbraedii*, and flowers are from *Mimulopsis arboreus*, which reproduces on a 7 year cycle. Our occasional observations of eating wood and bark fit no clear pattern.

(*f*) *Animals*

We have no records of predation on insects or other invertebrates, nor have we found invertebrate remains in dung samples from Kanyawara ($n = 839$), or Ngogo ($n = 416$), despite the abundant presence of a variety of potential prey species (e.g. *Dorylus* sp. and *Crematogaster* sp.). Chimpanzees ate honey from *Apis mellifera* nests on several occasions.

Vertebrates are eaten regularly. Chimpanzees were observed eating nine *Colobus badiums*, three *Colobus guereza*, one *Cercocebus albigena* and one *Cercopithecus mitis*. Vertebrate remains were found in 2.9% of dungs both at Kanyawara ($n = 839$) and at Ngogo ($n = 416$). Lower rates of vertebrate remains have been found in dung from the Mahale Mountains, Tanzania and Mount Assirik, Senegal (1.8% of 783, and 1.9% of 5777 respectively), and higher rates in Gombe, Tanzania (5.8% of 1963) (Wrangham & Riss 1990). This suggests that meat eating in Kibale occurs at intermediate rates compared with other chimpanzee populations, and that, as elsewhere, it provides occasional supplements to the diet rather than a daily source of high-quality food.

(*g*) *Geophagy*

Consumption of soil or termite clays was not observed.

4. BABOON DIET IN COMPARISON TO CHIMPANZEES

Like chimpanzees, baboons concentrated heavily on non-fig fruit trees in approximate proportion to fruit availability. However, during months when both baboon and chimpanzee dungs were collected (December 1989 to March 1991) baboons ate fewer figs than did chimpanzees. The overall percentage of baboon dungs containing fig seeds was low (66.7% at Ngogo, $n = 96$; 76.5% at Kanyawara, $n = 34$); and fig abundance scores were lower (baboons 1.1 ± 0.7, chimpanzees 2.0 ± 0.7, Wilcoxon $z = 2.61$, $p < 0.01$). In contrast, Ngogo baboons had high FBR scores (baboon monthly mean $= 2.34 \pm 0.91$), compared with Ngogo chimpanzees (2.01 ± 1.00) or Kanyawara chimpanzees (2.03 ± 0.75) (Wilcoxon $z = 1.79$, $p = 0.06$).

Ngogo baboon FBR scores were strongly correlated with Ngogo chimpanzee FBR scores ($r = 0.79$, d.f. $= 14$, $p < 0.001$), and only weakly with Kanyawara chimpanzee FBR ($r = 0.43$, d.f. $= 14$, n.s.). This suggests that monthly variation in FBR intake at Ngogo was influenced for chimpanzees and baboons by local environment factors.

5. VARIATION IN DIET WITH FRUIT ABUNDANCE

Fruit production was not correlated between figs and non-figs. Thus although FA indices which included non-figs were closely correlated with each other ($r^2 = 0.82$–0.93, d.f. $= 27$, $p < 0.001$), FA (fig) was correlated poorly with FA (all) ($r^2 = 0.37$, d.f. $= 27$, $p < 0.001$) and not at all with FA (non-fig) ($r^2 = 0.04$). Mean FA values were similar for non-figs (16.7 ± 17.5) and figs (13.5 ± 9.4), but variance among monthly FA values was greater for non-figs than figs (FA (non-figs), range $= 0$–80.8, $cv = 104.7\%$; FA (fig), range $= 1.8$–37.5, $cv = 69.9\%$; $F_{(27,27)} = 3.65$, $p < 0.01$; data log-transformed to ensure normality). Thus large peaks of fruit availability were primarily the result of extensive fruiting by non-figs, whereas figs represented a relatively consistent level of fruit production.

Variation in FA indices over time showed no clear relation to the annual cycle. FA (non-fig) peaked three times, in April 1988, October 1988 and November 1989. FA (fig) peaked in March 1988, September 1988 and February 1989. None of the FA indices was correlated with monthly rainfall ($r^2 = 0.00$–0.18, d.f. $= 27$, n.s.).

To find how the intake of fibrous foods was influenced by fruit availability, we used data from dung samples (these tests are one-tailed). There was no direct correlation between FBR and any FA index involving non-figs ($r^2 = 0.02$–0.09, d.f. $= 27$, p n.s.). However, fibre levels rose when FA (fig) was low ($r^2 = 0.15$, $p < 0.05$). There were no correlations between any FA index and GLF ($r^2 = 0.00$–0.01, d.f. $= 27$, n.s.). From these results only FA (fig) appears important, and the effect is small. Fibre intake levels were correlated across months among Ngogo chimpanzees and Ngogo baboons, suggesting that environmental factors also influence pith eating.

As piths are likely to be more nutritious during periods of growth, the effect of monthly rainfall was examined in pairwise combination with each FA index. In every multiple regression the combination of rainfall and food availability accounted for a significant proportion of variance in FBR scores ($0.01 < p < 0.02$). When the effect of rainfall was removed, the partial correlation of FA indices with FBR was significant and negative ($0.01 < p < 0.05$; $r = 0.39$–0.48). Conversely, the partial correlation of rainfall with FBR was significant and positive ($0.01 < p < 0.05$; $r = 0.56$–0.58) when the effect of rainfall was removed. The best FA predictor in the multiple regression was FA (all) ($r^2 = 0.39$, d.f. $= 2, 25$, $p < 0.01$). FA (non-fig) was the worst ($r^2 = 0.29$, d.f. $= 2, 25$, $p < 0.05$). Chimpanzees therefore increased their relative pith intake during periods when fruit was scarce, responding to both fig and non-fig fruit production. GLF scores, by contrast, showed no correlation with any FA indices.

The validity of summing FA indices from different species depends on the assumption that trees of the

Table 1. *Nutrient composition of major items eaten by chimpanzees*

(n, number of species contributing to data; RTDM, % room temperature dry matter; Other columns show % of dry matter for crude protein (CP), lipid (lip), water-soluble carbohydrates (WSC), hemicellulose plus cellulose (H+C), neutral-detergent fibre (NDF) and condensed tannin (CT). Data on 'other ape' foods are extracted from Calvert (1985), Hladik (1977), Malenky (1990), Rogers *et al.* (1990) and Watts (1984), by calculating means across food species within studies; figures show means of study means. Parts analysed are those eaten by apes, i.e. fruit pulp, etc.)

	n	RTDM	ash	CP	lip	WSC	H+C	NDF	CT
Kanyawara fruit	1	36.7	3.1	6.3	0.0	32.7	7.5	40.3	0.0
other ape fruit	23	26.7	3.0	7.7	1.7	38.6	32.2	64.6	6.2
Kanyawara leaf	4	24.8	—	24.1	0.8	3.0	31.5	41.5	1.5
other ape leaf	44	—	—	16.8	2.6	3.9	26.6	46.0	10.9
Kanyawara fig	9	—	7.7	7.9	3.5	12.6	—	35.6	0.5
other ape fig	2	17.9	—	3.5	3.4	32.4	—	—	12.9
Kanyawara pith	8	11.9	13.4	9.3	0.8	9.8	46.9	50.5	0.1
other ape pith	21	17.0	8.1	10.5	1.6	8.4	48.0	56.3	—

same basal area produce equivalent amounts of food. Although this is probably reasonable, we can avoid relying on it by testing the relation between fruit abundance and fibre intake for the most frequently eaten tree-fruit, *M. bagshawei*, which were eaten in approximate proportion to their availability (figure 1). In Kanyawara the longest fruiting season was from August 1989 to February 1990. During this period (including one month before and after the fruiting period) there was a negative correlation between FBR scores and the FA index for *M. bagshawei* ($r^2 = 0.55$, d.f. $= 8$, $p < 0.05$). The same relation held between FBR and the number of *M. bagshawei* seeds per 100 g of dung, both for Kanyawara ($r^2 = 0.68$, d.f. $= 8$, $p < 0.01$) and Ngogo ($r^2 = 0.71$, d.f. $= 8$, $p < 0.01$). This suggests that, both at Kanyawara and at Ngogo, chimpanzees reduced their intake of fibrous stems when more *M. bagshawei* fruit was available. Again, however, GLF scores were not correlated with the index of fruit abundance ($r^2 = 0.11$–0.31, d.f. $= 8$, n.s.).

6. NUTRIENT COMPOSITION OF MAJOR FOOD TYPES

Fruits eaten by apes and other primates tend to have high concentrations of sugars and low concentrations of protein (table 1), whereas leaves have low sugar and high protein values (Hladik 1977; Rogers *et al.* 1990). We have also previously shown that Kanyawara figs tend to have low protein values in the edible pulp and that they supply digestible calories at rates equivalent to non-fig fruits (Wrangham *et al.* 1993; also see table 1). Our data on fruit and leaves therefore conform to the principle that large-bodied frugivores eat fruits for energy and leaves for protein (e.g. Milton 1980).

As leaves and pith are often eaten together during TPL sessions, it might be thought that they have equivalent nutritional significance, i.e. that piths are also complementary to low-protein fruits. Yet we have shown that fibrous strands from piths were eaten more when fruit are scarce, whereas we have no evidence that leaf intake varied with fruit abundance. Possibly, therefore, piths provide an alternative energy supply to fruits. We therefore ask here whether piths tend to supply sugars, or protein, or other sources of nutrients.

We analysed nine species of herbaceous piths, including all the major genera eaten by chimpanzees in Kanyawara. Table 1 shows that the mean values from our samples are generally similar to the mean values for other species of pith eaten by apes.

First, protein concentrations (Kanyawara, mean 9.3%; other apes' mean, 10.5%) are similar to those in fruits (6.3–7.7%) and lower than those in leaves (16.8–24.1%) (Kanyawara data, Mann–Whitney $n_1 = 6$, $n_2 = 9$, $z = 2.83$, $p < 0.01$). This suggests that protein concentrations do not generally account for ape interest in pith. In support, table 2 shows that protein concentrations vary widely between pith species, from 1.7% (*Cyperus papyrus*) to 26.3% (*Acanthus pubescens*). Pith species eaten frequently include piths with both high protein (*A. pubescens*) and low protein (*P. purpureum*). Thus we find no consistency with regard to protein selection.

Concentrations of water-soluble carbohydrates in pith are greater than in leaves (table 1), although not significantly so (Mann–Whitney $n_1 = 6$, $n_2 = 9$, $z = 0.71$, n.s.). Species values again show substantial variation. For example, *P. purpureum* has almost ten times the sugar concentration of the two species of *Piper*. As with protein, therefore, sugar concentrations do not provide a consistent explanation for the selection of piths.

Milton & Demment (1988) showed that hemicellulose and cellulose were both partly digested by captive chimpanzees. We therefore examined values of these two fibre fractions to find out whether piths tend to have high concentrations of these fermentable components. Table 1 shows that the mean concentration of hemicellulose and cellulose in pith is indeed high (46.9% in Kanyawara, 48.0% for other ape samples) compared to both fruits (7.5%, 32.2%) and leaves (26.6%, 31.5%). We cannot yet test the difference between fruits and piths with our own data, but Calvert (1985) provided appropriate data from piths eaten by gorillas in Cameroon: piths had higher concentrations of hemicellulose and cellulose than fruits ($n_1 = 7$, $n_2 = 9$, $z = 1.96$, $p = 0.05$). The difference between pith and leaf concentrations was testable in two sets of samples: concentrations of cellulose plus hemicellulose were higher in pith than leaf both in the Kanyawara data ($n_1 = 6$, $n_2 = 9$, $z = 2.00$, $p < 0.05$)

Table 2. *Nutrient composition of herbaceous piths eaten by Kanyawara chimpanzees*

(Abbreviations as in table 1. H, hemicellulose; C, cellulose; NDF, neutral-detergent fibre; lig, lignin.)

	n	CP	WSC	H	C	H+C	NDF	lig
Acanthus pubsecens	3	26.3	14.6	14.2	17.1	31.3	32.6	8.2
Aframomum mildbraedii	2	6.3	5.8	20.1	33.1	53.2	59.7	6.5
A. zambesiacum	3	6.7	5.6	20.9	33.0	53.9	57.9	4.0
Cyperus papyrus	2	1.7	16.4	30.2	29.5	59.7	62.8	3.1
Marantochloa leucantha	3	7.9	15.2	20.2	31.8	52.0	54.5	2.5
Pennisetum purpureum	1	4.4	21.8	19.7	38.6	58.3	63.3	4.9
Piper capensis	1	12.3	2.4	8.8	18.6	27.4	33.7	3.3
P. umbellatum	1	9.6	3.5	12.5	17.9	30.4	30.9	3.5
Renealmia congolana	1	8.8	12.5	21.8	33.4	55.2	59.5	4.3

and in Calvert's (1985) data ($n_1 = 8$, $n_2 = 9$, $z = 2.79$, $p < 0.01$). This difference between piths and leaves was the result of lower lignin levels in piths (Kanyawara, $z = 2.47$, $p < 0.02$; Cameroon, $z = 2.21$, $p < 0.05$). Total fibre (i.e. neutral detergent fibre, NDF), by contrast, did not differ between piths and leaves (Kanyawara, $z = 1.59$, n.s.; Cameroon, $z = 1.34$, n.s.).

7. DISCUSSION

Fibre levels increased as a function of rainfall and decreased in relation to our measures of fruit availability. Furthermore, fibre levels showed a negative correlation with the availability of the most important fruit crop, *M. bagshawei*, during a long fruiting season, and with the number of *M. bagshawei* seeds in the dung both at Kanyawara and Ngogo. We conclude that, in Kibale Forest, chimpanzees tend to respond to tree-fruit shortages by increasing their intake of piths. We found no evidence that leaf intake increased in parallel.

When one considers piths as a single food group, which seems to be how chimpanzees utilize them, our nutritional data support previous conclusions by Calvert (1985) and Rogers *et al.* (1990) that the piths eaten by apes are important sources of energy. However, unlike previous studies we found no evidence that protein concentrations influenced this food group's selection. Nor did we find, in contrast to Rogers *et al.* (1990), that piths provide significant sugar levels: our samples agree instead with Calvert (1985) in having low lignin levels and high NDF. The result is that fermentable fibre is in high concentration. We therefore hypothesize that the importance of piths in the chimpanzee diet tends to result from their providing calories in the form of fermentable fibre.

Previous studies have proposed that terrestrial herbs provide important fallback foods for bonobos and gorillas (Badrian & Malenky 1984; Kano 1983; Kano & Mulavwa 1984; Rogers *et al.* 1988) and mandrills (*Mandrillus sphinx*) (Hoshino 1986). Our results suggest the same is true for Kibale chimpanzees and baboons. However, chimpanzees are not restricted to piths as fallback foods. They exploit a variety of low-quality foods, thereby allowing them access to drier habitats than are known for bonobos or gorillas.

Although we propose that, in general, piths and leaves have different significance, providing energy and protein respectively, this scheme may be too simple. There is some indication that piths of monocotyledons and dicotyledons differ. The only dicotyledons in table 2 are *A. pubescens*, *P. capensis* and *P. umbellatum*, each of which has low levels of hemicellulose and cellulose compared with the monocotyledons. Monocotyledons are predominant in the herbaceous diets of bonobos, forest gorillas, mandrills and chimpanzees, and they appear consistently to provide a critical resource for semi-terrestrial forest primates. Their value as a food source is presumably a result of their basal growth, which leads to a concentration of mobilized nutrients in a relatively compact volume. Thick, tough protective stems may prevent most animals from harvesting piths, leaving them primarily to apes and elephants.

Among hominoids thin-enamelled teeth are restricted to the African apes, the only species that rely heavily on terrestrial piths. Orangutans (*Pongo pygmaeus*), by contrast, have thick-enamelled teeth and only occasionally eat piths (Kay 1981; Leighton 1993), although they spend as much time eating leaves as do chimpanzees (Rodman 1977). This suggests that the divergence of African apes and hominoids, which was probably also a division into thin-enamelled and thick-enamelled groups, marked a divergence into pith eaters and non-pith eaters. The thick-enamelled Miocene apes can be expected to have had a different vital food resource from the thin-enamelled apes of today, with consequent differences in many ecological and social variables.

Given the ubiquity and abundance of terrestrial herbs in contemporary African forests, together with their apparent importance for apes and baboons, it is curious that pith eating is so much less important in other continents and, apparently, in other eras (i.e. in the Miocene). The issue is of interest partly because the occurrence of an essential food resource in the ground layer may have influenced the evolution of terrestriality. As a stimulus to further investigation, we speculate that African rain-forests may have higher biomass densities of edible herbaceous piths than other continents. If so, the causes of such differences may help explain the origins of the modern African apes.

We thank the Government of Uganda, and especially the Forest Department, for permission to work in the Kibale Forest Reserve. Facilities were provided by Makerere University Biological Field Station. The Department of Zoology, Makerere University, assisted at all times. Funding was generously provided by the National Science Foundation (BNS-8704458), National Geographic Society (3603-87),

MacArthur Foundation and Leakey Foundation. Assistance in fieldwork was given by F. Amanyire, H. Bagonza, J. Baptiste, J. Basigara, A. Clark, K. Clement, B. Gault, M. Hauser, G. Kagaba, C. Katongole, T. Lawrence, R. Marumba, C. Muruuli, P. Novelli, C. Opio, E. Tinkasimire and P. Tuhairwe. A. Katende, P. Ipulet and J. Kasenene kindly identified plants. Special thanks are due to A. Clark, G. Isabirye-Basuta, and A. Johns for support in the field.

REFERENCES

Association of Official Analytical Chemists (AOAC) 1984 Fat (crude) or ether extract in animal feeds: direct method. In *Official methods of analysis of the Association of Official Analytical Chemists* (ed. S. Williams), pp. 159–160. Arlington, Virginia: Association of Official Analytical Chemists.

Badrian, N. L. & Malenky, R. K. 1984 Feeding ecology of *Pan paniscus* in the Lomako Forest, Zaire. In *The pygmy chimpanzee: evolutionary biology and behaviour* (ed. R. L. Susman), pp. 275–299. New York: Plenum Press.

Badrian N., Badrian, A. & Susman, R. W. 1981 Preliminary observations on the feeding behaviour of *Pan paniscus* in the Lomako Forest of Central Zaire. *Primates* **22**, 173–181.

Bate-Smith, E. C. 1975 Phytochemistry of proanthocyanidins. *Phytochemistry* **14**, 1107–1113.

Berg, C. C. & Hijman, M. E. E. 1989 *Flora of topical East Africa: Moraceae*. Rotterdam: A. A. Balkema.

Butynski, T. M. 1990 Comparative ecology of blue monkeys (*Cercopithecus mitis*) in high- and low-density subpopulations. *Ecol. Monogr.* **60**, 1–26.

Calvert, J. J. 1985 Food selection by western gorillas (*G. g. gorilla*) in relation to food chemistry. *Oecologia, Berl.* **65**, 236–246.

Dubois, M., Gilles, K. A., Hamilton, J. K., Rebers, P. A. & Smith, F. 1956 Colorimetric methods for determination of sugars and related substances. *Analyt. Chem.* **28**, 350–356.

Ghiglieri, M. P. 1984 *The chimpanzees of Kibale Forest*. New York: Columbia University Press.

Goering, H. K. & van Soest, P. J. 1970 *Forage fiber analysis*. Agricultural Handbook No. 379. Washington D.C.: A.R.S., U.S.D.A.

Goodall, A. G. 1977 Feeding and ranging behaviour of a mountain gorilla group (*Gorilla gorilla beringei*) in the Tshibinda-Kahuzi region (Zaire). In *Primate ecology* (ed. T. H. Clutton-Brock), pp. 449–479. New York: Academic Press.

Hladik, C. M. 1977 Chimpanzees of Gabon and chimpanzees of Gombe: some comparative data on the diet. In *Primate ecology* (ed. T. H. Clutton-Brock), pp. 481–501. New York: Academic Press.

Hoshino, J. 1986 Feeding ecology of mandrills (*Mandrillus sphinx*) in Campo Animal Reserve, Cameroon. *Primates* **27**, 248–273.

Isabirye-Basuta, G. 1990 Feeding ecology of chimpanzees in the Kibale Forest, Uganda. In *Understanding chimpanzees* (ed. P. G. Heltne & L. A. Marquardt), pp. 116–127. Cambridge: Harvard University Press.

Kano, T. 1983 An ecological study of the pygmy chimpanzees (*Pan paniscus*) of Yalosidi, Republic of Zaire. *Int. J. Primatol.* **4**, 1–31.

Kano, T. & Mulavwa, M. 1984 Feeding ecology of the pygmy chimpanzees (*Pan paniscus*) of Wamba. In *The pygmy chimpanzee: evolutionary biology and behaviour* (ed. R. L. Susman), pp. 233–274. New York: Plenum Press.

Kay, R. F. 1981 The nut-crackers – a new theory of the adaptations of the Ramapithecinae. *Am. J. phys. Anthrop.* **55**, 141–151.

Leighton, M. 1993 Modelling diet selectivity by Bornean orangutans: evidence for integration of multiple criteria in fruit selection. *Int. J. Primatol.* **14**, (In the press.)

Malenky, R. K. 1990 Ecological factors affecting food choice and social organization in *Pan paniscus*. Ph.D. dissertation, SUNY, Stony Brook.

McGrew, W. C., Baldwin, P. J. & Tutin, C. E. G. 1981 Chimpanzees in a hot, dry and open habitat: Mt. Assirik, Senegal, West Africa. *J. hum. Evol.* **10**, 227–244.

McGrew, W. C., Baldwin, P. J. & Tutin, C. E. G. 1988 Diet of wild chimpanzees (*Pan troglodytes verus*) at Mt. Assirik, Senegal: I. Composition. *Am. J. Primatol.* **16**, 213–226.

Milton, K. 1980 *The foraging strategies of howler monkeys: a study in primate economics*. New York: Columbia Press.

Milton, K. & Demment, M. W. 1988 Digestion and passage kinetics of chimpanzees fed high and low fiber diets and comparison with human diets. *J. Nutr.* **118**, 1082–1088.

Mole, S. & Waterman, P. G. 1987 A critical analysis of techniques for measuring tannins in ecological studies: I. Techniques for chemically defining tannins. *Oecologia, Berl.* **72**, 137–147.

Nishida, T. 1976 The bark-eating habits in primates, with special reference to their status in the diet of wild chimpanzees. *Folia primat.* **25**, 277–287.

Nishida, T. 1979 The social structure of chimpanzees of the Mahale mountains. In *The Great Apes* (ed D. A. Hamburg & E. R. McCown) pp. 72–121. Menlo Park: Benjamin/Cummings.

Nishida, T. 1989 A note on the chimpanzee ecology of the Ugalla area, Tanzania. *Primates* **30**, 129–138.

Nishida, T. & Uehara, S. 1983 Natural diet of chimpanzees (*Pan troglodytes schweinfurthii*): long-term record from the Mahale Mountains, Tanzania. *Afr. Stud. Monogr.* **3**, 109–130.

Pierce, W. C. & Haenisch, E. L. 1947 *Quantitative analysis*, 2nd edn. London: John Wiley & Sons.

Robertson, J. B. & van Soest, P. J. 1980 The detergent system of analysis and its application to human foods. In *The analysis of dietary fiber in foods*, (ed. W. P. T. James & O. Theander), pp. 123–158. New York: Marcel Dekker.

Rodman, P. S. 1977 Feeding behaviour of orang-utans of the Kutai Nature Reserve, East Kalimantan. In *Primate ecology* (ed. T. H. Clutton-Brock), pp. 381–413. London: Academic Press.

Rogers, M. E. & Williamson, E. A. 1987 Density of herbaceous plants eaten by gorillas in Gabon: some preliminary data. *Biotropica* **19**, 278–281.

Rogers, M. E., Williamson, E. A., Tutin C. E. G. & Fernandez, M. 1988 Effects of the dry season on gorilla diet in Gabon. *Primate Reports* **22**, 25–33.

Rogers, M. E., Maisels, F., Williamson, E. A., Fernandez, M. & Tutin, C. E. G. 1990 Gorilla diet in the Lopé Reserve, Gabon: a nutritional analysis. *Oecologia, Berl.* **84**, 326–339.

Strickland, J. D. H. & Parsons, T. R. 1972 *A practical handbook of seawater analysis*. Ottawa: Fisheries Board of Canada.

Suzuki, A. 1969 An ecological study of chimpanzees in a savanna woodland. *Primates* **10**, 103–148.

van Soest, P. J. & Robertson, J. B. 1991 Analysis of forages and fibrous foods, a laboratory manual.

Watts, D. P. 1984 Composition and variability of mountain gorilla diets in the central Virungas. *Am. J. Primatol.* **7**, 323–356.

Wrangham, R. W. 1977 Feeding behaviour of chimpanzees in Gombe National Park, Tanzania. In *Primate ecology* (ed. T. H. Clutton-Brock), pp. 503–538. London: Academic Press.

Wrangham, R. W. 1986 Ecology and social relationships in

two species of chimpanzee. In *Ecological aspects of social evolution* (ed. D. I. Rubenstein & R. W. Wrangham), pp. 325–378. Princeton University Press.

Wrangham, R. W. & Riss, E. V. Z. B. 1990 Rates of predation on mammals by Gombe chimpanzees, 1972–1975. *Primates* **31**, 157–170.

Wrangham, R. W., Clark, A. P. & Isabirye–Basuta, G. 1992 Female social relationships and social organization of Kibale Forest chimpanzees. In *Human origins* (ed. T. Nishida, W. C. McGrew, P. Marler, M. Pickford & F. de Waal). University of Tokyo Press. (In the press.)

Wrangham, R. W., Conklin, N. L., Etot, G., Obua, J., Hunt, K. D., Hauser, M. D. & Clark, A. P. 1993 The value of figs to chimpanzees. *Int. J. Primatol.* **14**. (In the press.)

Discussion

K. Milton (*Department of Anthropology, University of California, Berkeley, U.S.A.*). I suggest that considerable pith may not be swallowed, as with bats and spider monkeys eating fig fruit. Thus chimpanzees may take in more soluble carbohydrates that might be apparent, whilst sparing their digestive tracts from having to process much of the fibrous matter.

N. L. Conklin. So far it is not known what percentage of pith is wadged as opposed to swallowed. Habituating the chimps to letting us follow them on the ground has been a very recent accomplishment. Wadging is definitely something the authors plan to investigate more thoroughly now that they are getting better terrestrial observations.

E. Rogers (*Scottish Primate Research Group and Institute of Cell, Animal and Population Biology, University of Edinburgh, U.K.*). Single measures of sugar content in piths may not give the whole story. There are probably variations between individuals, and in different seasons, so it may be misleading to assume piths remain low in sugar content throughout the year. Species with higher sugar concentrations (20% dry mass in their pith, which have been found in Kibale and Lope, suggest that apes may obtain water-soluble sugars from pith some of the time.

A. Whiten (*Scottish Primate Research Group, University of St. Andrews, U.K.*). The authors seem to be attempting to find a single factor which is 'the' explanation of why a particular food type like pith is eaten. There should be caution about assuming such single factors; might it not be that different piths are eaten for different nutritional reasons, and that in some or all cases, the explanation has to do with more than one component? In the case of two pith species at least (*Acanthus pubescens* and *Piper umbellatum*, see table 2), protein levels are quite high and these are the ones with relatively low hemicellulose and cellulose levels; maybe it is the combined contribution of these which makes these piths valuable?

N. L. Conklin. I agree that the pith picture may be more complicated than presented here. The reason the authors chose to look at all piths as one group is because of an aspect of pith eating that was not mentioned in the paper. It appears that on most afternoons, chimps will leave the fruiting tree they have been feeding in, come down to the ground and eat pith for a few minutes to half an hour. So far there are not enough data to show patterns to the pith species chosen for consumption each day and how that relates to what else has been eaten that day.

The two high protein piths are actually the least frequently eaten as far as we can tell so far. Pennisetum is one of the most frequently eaten, as well as readily available. Aframomum and Marantochloa are also frequently eaten and common. It seems that when they eat pith, they feed on one species per session. The two high protein piths are less common in the environment but the authors have not looked at selectivity among piths yet. Richard Wrangham hopes to find a Ugandan masters student to pursue the pith question further.

R. A. Barton (*University of Sheffield, U.K.*). The authors mentioned that baboons are also known to eat some of these fibrous foods exploited by the great apes. Baboons are, of course, much smaller than apes, and hence presumably less capable of digesting high-fibre foods. Is there any evidence that the baboons are in some way more constrained than the apes in their use of fibrous foods as a fall-back? If so, does this mean that the habitat is something of a marginal one for baboons, which might be indicated by low population density and long day journeys relative to group size, or do they have some other way of surviving hard times?

N. L. Conklin. The forest baboons in Kibale have been very little studied. As far as we can tell they are very migratory and the habitat seems to be a marginal one for them but their home range is not known. They do wadge fibrous foods, which is one way to get out the soluble nutrients without having to ingest and pass the fibre, but then chimps wadge pith and fruit also. From information obtained at other sites it is known that baboons will eat plant material higher in secondary plant compounds than do chimps, but during times of food shortage in Kibale the baboons tend to disappear from the study site and it is not known where they go or what they do.

Foraging profiles of sympatric lowland gorillas and chimpanzees in the Lopé Reserve, Gabon

CAROLINE E. G. TUTIN[1]†, MICHEL FERNANDEZ[1], M. ELIZABETH ROGERS[2], ELIZABETH A. WILLIAMSON[3] AND WILLIAM C. McGREW[3]

[1] *Centre International de Recherches Médicales de Franceville, Gabon; Scottish Primate Research Group, University of Stirling, U.K.*
[2] *Scottish Primate Research Group and Institute of Cell, Animal and Population Biology, University of Edinburgh, U.K.*
[3] *Scottish Primate Research Group and Department of Psychology, University of Stirling, U.K.*

SUMMARY

Comparison of the diets of sympatric gorillas and chimpanzees allows an analysis of niche separation between these two closely related species. Qualitatively, their diets are similar, being dominated by an equally diverse array of fruit species complemented with vegetative plant parts, seeds and insects. Gorillas eat more vegetative plant parts than do chimpanzees, but niche separation is most obvious in periods of fruit scarcity when the two species show different strategies that reduce competition for food. Their abilities to overcome mechanical and physical plant defences appear to differ, as gorillas are able to subsist entirely on abundant vegetative foods. Chimpanzees show social adjustment, foraging alone or in small groups, to reduce intra-specific competition for scarce fruit resources. Thus it seems that subtle physiological differences have far-reaching repercussions, defining potential evolutionary pathways for social organization and allowing sufficient niche separation between species.

1. INTRODUCTION

Foraging profiles describe the ways in which a species searches for food, what foods it selects, and the strategies it uses in competing for food, both with members of its own species and with those of sympatric species with overlapping diets. Complete description of the diets of wild primates is not easy, even in the simplest terms of compiling a qualitative list of species and parts eaten. The number of food items recorded for a species varies both with the duration of study and with habitat. The habitat defines the potential diet of each species, and floristic variation is often so great as to limit inter-site comparisons. Food availability changes seasonally, and dramatic inter-annual variations have been recorded in most long-term studies. However, the distribution of different plant food categories (fruits, leaves, insects, etc.) is consistent within most primate species (Sussman 1987), allowing classification of diet as frugivorous, folivorous or insectivorous. Morphological and physiological adaptations to diet are often clear, and multi-factorial allometric analyses have shown that the dietary class correlates with some physical and social variables (Clutton-Brock & Harvey 1980).

With respect to the African apes, chimpanzees (*Pan troglodytes*) have been classed as frugivores, and gorillas (*Gorilla gorilla*) as folivores. For all populations of chimpanzees studied to date, fruit dominates the diet, both qualitatively in percentage terms of total foods and quantitatively in terms of percentage of feeding time and mass ingested (Hladik 1973; Wrangham 1977). Gorilla diet appears to be determined by the variety of foods available in the habitat, with the proportion of fruit increasing with plant species diversity: lowest in the montane habitat of mountain gorillas, intermediate in disturbed forest habitats and highest in lowland tropical forest (Tutin & Fernandez 1985).

Dietary niche is an important ecological constraint on the evolution of primate social systems (Terborgh 1983; Wrangham 1987), as different foods are dispersed differently in time and space, defining the extent and nature of intra-specific competition for food. Allometric analyses suggest a relation between brain size (relative to body mass) and diet, with frugivores having relatively larger brains (Harvey & Clutton-Brock 1983). Maintaining a frugivorous diet presents huge intellectual challenges of memory and spatial mapping compared with the relative ease of harvesting abundant foliage foods (Milton 1981).

The theory of competitive exclusion predicts clear niche differentiation and differences in foraging profiles between two species as closely related and morphologically similar as the gorilla and chimpanzee. Gorillas and chimpanzees are similar genetically (Sarich 1983), in gut morphology (Chivers & Hladik 1984), craniodental anatomy (Shea 1983), gut passage time (Milton 1984) and longevity, but there are differences in social organization, sexual dimorphism, and intestinal flora and fauna. For example, gorillas have cohesive groups led usually by a single fully adult male, whereas chimpanzees live in large fission–fusion communities.

† All correspondence to: Dr C. Tutin, C.I.R.M.F., B.P. 769, Franceville, Gabon.

Sexual dimorphism is pronounced in gorillas, with mature males weighing twice as much as females, but is less marked in chimpanzees, with adult females weighing about three-quarters as much as adult males (Groves 1970; Wrangham & Smuts 1980). The intestinal flora and fauna are ubiquitous and more diverse in wild gorillas than in chimpanzees (Ashford *et al.* 1990; File *et al.* 1976; Goussard *et al.* 1983).

Meaningful comparison of diet and foraging strategies can best be made by using the same methods to study sympatric populations. This is especially true when the two species are closely related, as the presence of one is likely to influence the other. Here we present seven years of data to compare the diets of sympatric western lowland gorillas (*Gorilla g. gorilla*) and chimpanzees (*Pan t. troglodytes*) in tropical rain forest in the Lopé Reserve, Gabon.

2. STUDY SITE AND METHODS

The study area covers approximately 40 km^2 of tropical rain forest in the Lopé Reserve (0° 10′ S, 11° 35′ E) in central Gabon. Chimpanzees and gorillas occur at similar population densities of about one individual per square kilometre. Average annual rainfall is 1536 mm (1984–1990) and the climate is characterized by a three month dry season from mid-June to mid-September.

Research on the gorillas and chimpanzees began in late 1983 and is ongoing. Field procedure involves searching through the forest for gorillas and chimpanzees, or for indirect signs of their presence and activities. At the beginning of the study the apes were unused to humans and usually fled when encountered. By 1988, two groups of gorillas had become partly habituated, but chimpanzees still remain shy and, as visibility is generally poor, observation is limited. Thus, in addition to observation, we use indirect methods to describe diet, systematically collecting fresh faeces and describing feeding remains.

Faecal samples collected in the field are sealed in plastic bags and later weighed and washed in sieves with 1 mm mesh. The particulate remains are examined macroscopically and the contents listed, large seeds counted and small seeds rated on a four-point scale of abundance (abundant, common, few, rare). Non-fruit plant parts (leaves, stems, pith and bark) are represented in faeces by fibre and partly digested fragments of leaves. These cannot be identified to species level macroscopically, so the volume of the categories 'fibre' and 'green leaf fragments' is assessed with respect to the total mass of the faecal sample and rated on the same four-point abundance scale. We computed a numerical value for the proportion of non-fruit remains in faeces by converting the abundance ratings as follows: abundant = 4; common = 3; few = 2; rare = 1. The combined score of all non-fruit categories gives the 'foliage score' and faeces were classed as 'foliage dominated' if the foliage score was 4 or more.

Phenological data on leaf, flower and fruit production have been collected monthly since 1984. Initially, data were collected on five individuals of 83 common tree species, but since October 1986 we have monitored ten individuals of each of 60 species of woody plants (56 trees, three shrubs and one liana), the fruit of which is important in the diet of gorillas and chimpanzees. Each tree is examined at the beginning of each month, and the relative quantities of flowers, fruit (immature and ripe) and leaves (flush, mature and senescent) are scored on a ten-point scale.

The data come from seven years of continuous research at Lopé (January 1984–December 1990), during which we have had 653 contacts with gorillas and 791 with chimpanzees. At least eight groups of gorillas and two communities of chimpanzees have been involved.

3. RESULTS

(*a*) *Composition of diet*

Qualitatively, the diets of gorillas and chimpanzees at the Lopé are similar. To date we have identified 142 chimpanzee foods and 203 gorilla foods. Table 1 compares the diet in terms of the different food types eaten. Fruit is the most numerous food class for both apes: gorillas eat fruit of at least 91 species and chimpanzees of at least 96 species. Dietary overlap is extensive; only 4% of gorilla fruits and 7% of chimpanzee fruits are not shared.

Figure 1 compares the diversity of fruit species eaten by gorillas and chimpanzees using data from five randomly selected faecal samples per species per month over a three-year period. The plots are similar with the number of different species of fruit eaten by gorillas equal to, or slightly greater than, those eaten by chimpanzees. Remains of at least one species of fruit were found in 96% of the 4301 gorilla faeces and in 98% of the 1656 chimpanzee faeces analysed over seven years. The mean number of different fruit species per faecal sample was 3.0 (range 0–10) for gorillas and 2.7 (range 0–9) for chimpanzees.

Both gorillas and chimpanzees regularly consume insects. Remains of at least one species of insect were found in 30% of gorilla faeces and 31% of chimpanzee faeces. The species eaten most often by both apes was the weaver ant (*Oecophylla longinoda*), but there was little overlap in the other species eaten. Chimpanzees used tools to obtain two large species of social ants and the honey of *Apis* sp., whereas gorillas ate three or more species of small ants.

(*b*) *Seasonal variation in diet*

The range of ripe fruit available to the apes at Lopé varied considerably over the annual cycle. Figure 2 shows the average number of species in the phenology sample (total $n = 60$, including 53 species eaten by gorillas and 57 by chimpanzees) bearing ripe fruit in each month over a four-year period (October 1986–September 1990). Few (4–7) species in the phenology sample bear ripe fruit during the long dry season from June to August, and the maximum diversity of fruit (16–19 species) is available from November to January.

Table 2 shows how three measures of diet vary over

Table 1. *Comparison of diets of gorillas and chimpanzees at Lopé*

	gorilla		chimpanzee	
category of food	number of species eaten	percentage of total	number of species eaten	percentage of total
fruit	91	44.8	96	67.6
young leaves	27	13.3	14	9.9
mature leaves	22	10.8	2	1.4
immature seeds	11	5.4	4	2.8
ripe seeds	10	4.9	6	4.2
pith	17	8.4	3	2.1
bark	9	4.4	1	0.7
flowers	3	1.5	3	2.1
miscellaneous[a]	8	3.9	2	1.4
insects	5	2.5	8	5.6
mammals	0	0.0	3	2.1
total	203	99.9	142	99.9

[a] Includes roots, galls and fungi.

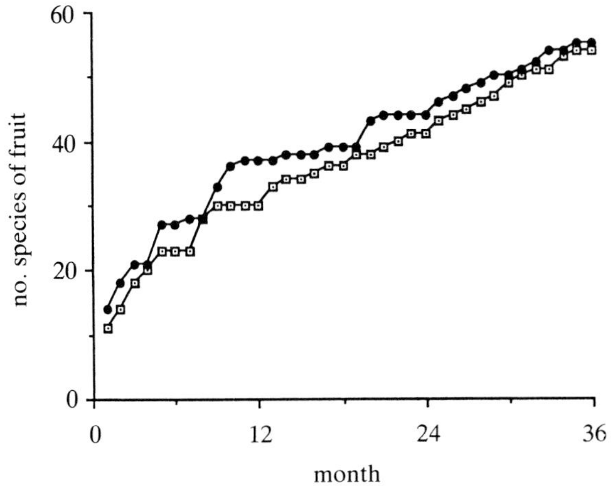

Figure 1. Cumulative number of fruit species recorded in five faecal samples per month (squares, chimpanzee; circles, gorilla).

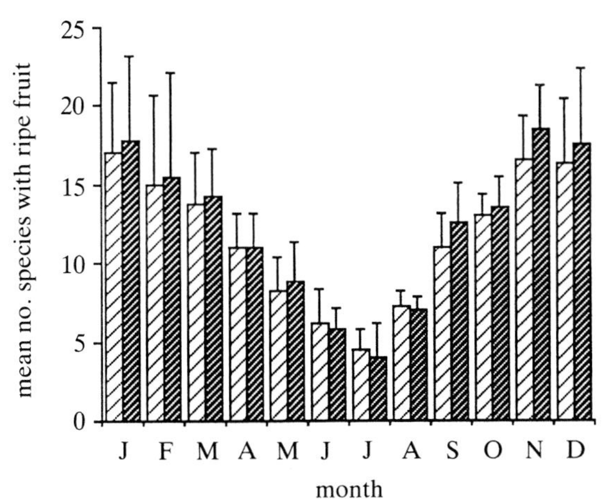

Figure 2. Phenology data: mean number of species with ripe fruit eaten by gorillas (light shading) and by chimpanzees (bold shading), 1986–1990.

the year. The mean number of different fruit species found per faecal sample reflects the diversity of the fruit diet: for gorillas, the maximum of 4.2 fell in January and the minimum of 1.6 in July; for chimpanzees, the maximum of 3.2 was recorded in March and April and the minimum of 1.5 was in July. The annual variation correlates positively with that of the fruit phenology data, reaching significance for gorillas (Spearman Rank Correlation, $r_s = 0.61$, $n = 12$, $p < 0.05$) but not for chimpanzees ($r_s = 0.55$, $n = 12$, $p = 0.06$). Although data from faecal analysis allow quantification of the diversity of the fruit part of the diet, they do not easily allow estimation of the absolute quantities of fruit ingested. Instead, we used foliage scores (see §2) to compare the relative amounts of non-fruit foods in faecal samples. Foliage scores for gorillas were consistently higher than those for chimpanzees, but annual variation was similar, with maximum values during the long dry season, June to August. Annual variation in the mean number of different fruit species recorded in the total monthly sample of faeces was much less marked, but minimum values for both gorillas and chimpanzees fell in June. Gorillas ate a wider variety of fruit than chimpanzees in eight months of the year.

The frequency of insectivory showed no significant correlation with the diversity of fruit eaten or with foliage scores, but chimpanzees did eat most insects in the first two months of the long dry season when their foliage scores were highest and fruit diet least diverse.

(*c*) *Inter-annual variation*

Inter-annual variation in fruit food diversity was considerable. No clear correlation emerged between fruit abundance and rainfall, and it is probable that differences in plant productivity were caused by a number of factors, including inherent species-specific, non-annual rhythms, insect attacks, and climate. Figure 3 compares the number of species in the phenology sample with fruit eaten by gorillas that had ripe fruit in 1987–1988 and 1989–1990, respectively the 'worst' and 'best' fruit years since expanded data collection began in October 1986 (see §2). Data are presented from August to July, as the climatic break of the long dry season appears to affect productivity rhythms over the ensuing 12 months.

Figure 3 shows that in five months of the 'good' fruit

Table 2. *Seasonal variation in chimpanzee and gorilla diet*

	mean fruit species per faecal sample		mean foliage score		total fruit species eaten per month	
month	chimpanzee	gorilla	chimpanzee	gorilla	chimpanzee	gorilla
January	2.9	4.4	2.6	4.5	12.3	17.6
February	3.1	4.2	2.8	4.1	11.2	17.6
March	3.2	3.3	2.5	4.7	10.3	15.0
April	3.2	3.0	3.2	5.9	13.2	11.0
May	2.5	3.3	3.0	5.3	10.7	10.2
June	1.7	2.0	5.0	6.7	9.3	8.9
July	1.5	1.6	5.4	7.5	10.0	11.7
August	2.5	2.4	4.5	7.2	14.5	14.3
September	2.1	2.8	3.0	6.3	12.0	17.6
October	3.1	3.6	2.6	5.7	12.8	17.3
November	2.7	2.6	2.6	6.0	10.8	15.0
December	2.3	2.7	2.8	6.0	11.4	14.0

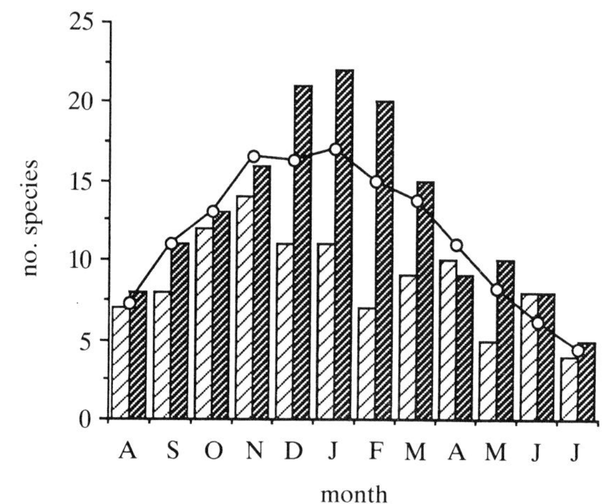

Figure 3. Number of species bearing ripe fruit eaten by gorillas (light shading, 1987–1988; bold shading, 1989–1990; circles, 5-year mean).

year, 1989–1990, fruit diversity was at least twice as great as in the 'bad' fruit year, 1987–1988, in which fruit diversity was low in December–May but not in October–November, a period when production is relatively constant (see figure 2, standard deviations). The plant phenology sample includes all of the major ape foods, except the oil palm, *Elaeis guineensis* (as the fruits are hidden from view), but does not take account of the density of the various tree species in the study area. The phenology data record only fruit diversity, not absolute abundance, but there was a significant negative correlation between the diversity of ripe fruit in the phenology sample each month and the average monthly foliage score from faeces over the two years (for gorillas, $r_s = -0.631$, $n = 24$, $p < 0.001$; for chimpanzees, $r_s = -0.641$, $n = 24$, $p < 0.001$), suggesting that the number of phenology species with ripe fruit does reflect differences in overall abundance of fruit foods.

Table 3 compares the diets of gorillas and chimpanzees in the 'bad' fruit year and the 'good' fruit year. The differences in the average number of different species of fruit per faecal sample are greater between years than between species. The comparison of foliage scores shows a similar general result but the relation is not as clear-cut. Chimpanzees only showed high foliage scores when fruit diversity was very low: in the dry season (July–August) and in five additional months in the 'bad' fruit year. Gorillas showed low average foliage scores (< 4) in only five of the 24 months (January–May 1990). In three of these months, diversity of ripe fruit was high (> 15 species) but this was not the case in April–May. Also, foliage scores were not low in November–December 1989 when fruit diversity was high (see figure 3). The percentage of faeces per month that were 'foliage dominated' (see §2) show that gorillas consume more foliage than chimpanzees, both species eat more foliage when fruit is scarce, and when fruit is abundant, gorillas continue to eat relatively large amounts of foliage.

(*d*) *Diet during periods of fruit scarcity*

During the three month dry season at the Lopé, gorillas eat large quantities of vegetative foods, some of which (leaves and bark of the large tree *Chlorophora excelsa*, and pith and young leaves of two species of herbs, *Marantochloa* spp.) are permanently available but are usually ignored. Chimpanzees increase their consumption of the fruit of *Elaeis guineensis* and of the pith and young leaves of herbaceous plants (Zingiberaceae and Marantaceae). Chimpanzees eat *Elaeis* fruit throughout the year (22 months out of 24 in the two years) but gorillas never eat it. *Duboscia macrocarpa* fruit are also available throughout the year as trees fruit asynchronously. Gorillas eat *Duboscia* fruit regularly (21 out of 24 months) but chimpanzees consume it only when other fruit is scarce (the dry season every year, and in April and May 1988).

In periods of fruit scarcity during 'bad' years, gorillas and chimpanzees both increase consumption of pith and young leaves of Zingiberaceae and Marantaceae and the fruit of *Duboscia*, but other foods, especially immature and ripe seeds of a range of species, are also eaten in large quantities. Gorillas eat *Chlorophora* and *Marantochloa* only rarely at such times, suggesting that these dry season keystone foods are the least preferred and eaten only when all other foods are unavailable.

Table 3. *Comparison of ape diets in a 'bad' (1987–1988) and a 'good' (1989–1990) fruit year*

	mean fruit spp. per faecal sample				mean foliage score				percentage of foliage dominated			
	chimpanzee		gorilla		chimpanzee		gorilla		chimpanzee		gorilla	
month	bad	good	bad	good	bad	good	bad	good	bad	good	bad	good
August	2.1	2.5	1.6	2.2	5.0	4.9	7.3	7.3	75	77	100	100
September	1.6	2.5	2.2	3.3	4.0	2.0	6.1	6.8	60	16	100	98
October	2.8	3.9	3.0	4.6	2.7	2.4	6.1	5.2	36	22	100	87
November	1.8	4.4	2.1	3.0	2.5	2.6	6.1	5.3	20	40	96	91
December	2.1	2.7	2.6	2.6	2.6	2.6	5.9	5.4	33	36	100	91
January	3.3	2.5	3.8	5.1	1.8	3.2	4.4	3.5	28	50	71	49
February	3.0	3.0	3.4	4.8	3.0	3.3	5.5	3.1	35	50	92	47
March	2.1	3.4	1.6	3.6	4.7	1.4	7.4	3.1	42	17	100	45
April	1.8	3.7	0.8	3.4	5.3	1.9	7.5	3.9	100	13	100	53
May	1.4	2.6	1.2	3.4	5.9	1.6	7.1	4.0	89	21	100	57
June	0.7	2.0	0.3	2.7	7.2	2.5	7.6	5.3	100	34	100	73
July	1.6	1.3	1.7	1.5	6.0	5.7	7.5	7.3	100	80	100	100

4. DISCUSSION

The diets of the two species of ape at Lopé are generally similar, being numerically dominated by fruit, with leaves as the second most important part (table 1). Although the fruit diet is similar, the diversity of fruit consumption by gorillas at Lopé somewhat exceeds that of chimpanzees (figure 1, table 2), showing that this population of lowland gorillas cannot accurately be described as folivorous. Both gorillas and chimpanzees at Lopé are generalized, opportunistic frugivores, but neither species subsists entirely on fruit. Insects are eaten at similar frequencies, and vegetative plant parts such as leaves and pith are eaten regularly, but in larger quantities by gorillas than by chimpanzees (table 2). When fruit is scarce, during the dry season each year and for longer periods in some years (figures 2 and 3), both chimpanzees and gorillas increase their foliage intake, but both continue to find a surprising diversity of fruit (table 3).

The fruits eaten by apes at Lopé are diverse in terms of size, colour and chemistry, but it is instructive to look in more detail at the few species of fruit that are eaten exclusively, or in much greater quantities, by one of the two species of ape. There are seven important fruit foods of chimpanzees that are not eaten by gorillas at Lopé: five of these, including *Elaeis*, have an 'oily' flesh high in lipids (Rogers *et al.* 1990). Fruit and seeds eaten by gorillas at Lopé are relatively low in lipids, suggesting a general avoidance of foods with high lipid content (Rogers *et al.* 1990). The small number of species of mature leaves eaten by chimpanzees (table 1) suggests general avoidance of foods with high fibre content or with secondary compounds such as tannins and phenolics.

Studies of frugivorous communities elsewhere suggest that dietary divergence is highest when preferred food (succulent fruit) is scarce, and that niche separation is clear only at such times (Gautier–Hion & Gautier 1979; Terborgh 1983). Terborgh (1986) suggested that frugivorous vertebrates survive periods of fruit scarcity by relying on a few plant foods that either produce fruit outwith the community peak (e.g. figs), or have long-lasting, protected fruit (e.g. palms). These keystone resources are exploited in a specialized way by different members of the frugivore community, reducing inter-specific competition for food.

The diets of gorillas and chimpanzees differ when preferred foods are scarce, with each species of ape having some exclusive keystone resources whilst sharing others. Diet during the dry season at Lopé shows only minor inter-annual variation. By definition, keystone resources must be available in significant quantities when succulent fruit is scarce. This is true of mature leaves, pith, and bark of common species, but young leaves, fruit, and seeds can only be dependable keystone foods if produced continuously on a community basis, as is the case for fruit of *Duboscia* and *Elaeis* and young leaves of *Chlorophora* at Lopé (C. E. G. Tutin, unpublished data).

Plant chemistry is also relevant to niche separation. Fibre and phenolics help to protect plants from herbivores as they decrease digestibility (Freeland & Janzen 1974), but fibrous foods can provide energy if the cellulose can be digested (Gaulin & Konner 1977). The entodiniomorph ciliates, ubiquitous in faeces of lowland gorillas in Gabon (Goussard *et al.* 1983), are symbiotic cellulose digesters (Collet *et al.* 1984). Chimpanzees also have ciliates, but they are less frequent in faecal samples (75% of samples compared with 100% for gorillas) and less diverse (two species for chimpanzees, at least four for gorillas) (Imai *et al.* 1991; File *et al.* 1976). Gorillas can survive on a totally folivorous diet but it appears that chimpanzees cannot, as, whereas foliage intake increases when fruit is scarce, they continue to eat more fruit than gorillas during the dry season (table 3). It is not clear why gorillas avoid high-lipid foods, or how they cope with foods with high tannin content.

Food choice is complex, and although the chemical and mechanical properties of plants and the morphological and physiological adaptations of herbivores have significant roles, other, often subtle, factors are also involved. Some social insects are accessible only if tools are used, and are eaten by chimpanzees but not gorillas. Another example is the keystone resource of gorillas, the aquatic herbs (*Marantochloa* spp.) that grow in marshes and along the banks of streams: gorillas harvest the pith and young leaves of these plants while wading in water up to 50 cm deep. Chimpanzees may not eat *Marantochloa* because they have a strong dislike of water, crossing even shallow streams through overhead trees or on bridges. However, explanations for other instances of dietary variance are less clear. Strong preferences for certain fruits also exist, which explains why foliage scores are not consistently related to fruit diversity (figure 3, table 3). When *Dialium* fruit are abundant they dominate the diets of both apes, perhaps because both the sugary mesocarp and some seeds are eaten, providing a balance of nutrients rarely available at a single feeding site. Although often rich in protein, seeds can be small, and lengthy processing is needed to remove them selectively. This may explain why seeds are largely ignored except when fruit is scarce.

If the key to niche separation of the sympatric gorillas and chimpanzees at Lopé is the ability of gorillas to survive lengthy periods of fruit scarcity on an almost entirely folivorous diet, there are important repercussions. Non-reproductive parts of both woody and herbaceous plants are widely available, both in space and time, compared with fruit. Abundant food is available for a group of gorillas both in a *Marantochloa* marsh and the crown of a flushing *Chlorophora*, but the number of ripe fruit on a single *Elaeis* is very limited. Thus within-species feeding competition for keystone resources is likely to be low for gorillas and high for chimpanzees. Chimpanzees make a social adjustment and forage alone, or in small subgroups, when fruit is scarce (Wrangham 1977), whereas gorillas always forage in the company of conspecifics. The degree of intra-specific competition for food, particularly between adult females, is thought to have been a major selective force on the evolution of ape social systems (Wrangham 1979, 1987). Data from Lopé support this but emphasize that, in tropical forest habitats, competition for preferred food is only critical for a minority of the time (see also Gautier–Hion & Gautier 1979; Terborgh 1983). A high overall degree of dietary overlap is to be expected between closely related species that share physical adaptations, such as craniodental anatomy and gut morphology. Gorillas and chimpanzees are remarkably similar in these respects (Chivers & Hladik 1984; Milton 1984; Shea 1983), but the difference in body size could be important in allowing greater consumption of fibre by gorillas.

Apes have large brains, and the challenge of finding and exploiting keystone food resources, as well as keeping track of widely dispersed fruit foods, may have been an important selective pressure in the evolution of intelligence. Regular shortages of preferred foods affect most, if not all, populations of frugivores, and must provide a powerful stimulus for innovative and inventive behaviours. Chimpanzee tool-use gives access to otherwise inaccessible foods, and many tool-use behaviours show cultural variation (Nishida 1987). The kernels of large, well-protected nuts provide a high-quality food, but the essential tool-use technology exists in only some of the chimpanzee populations with access to the resource. Chimpanzees at Lopé do not crack nuts with tools, but if they did, intra-specific feeding competition during the dry season might be reduced, because if *Elaeis* kernels were exploited as well as the flesh, more food would be available.

The fission–fusion social organization of chimpanzees allows rapid adjustment to different levels of food abundance and responds to individual needs and preferences (Wrangham 1979). Living in groups conveys advantages such as reducing the threat of predation (Alexander 1973), if the obstacle of competing for food with conspecifics can be overcome. Predation may not be a constant threat for gorillas and chimpanzees but can assume importance in some habitats (Tutin *et al.* 1981) or on rare occasions (Tsukahara & Nishida 1990), and the risk is likely to be greater for lone individuals (Hiraiwa–Hasegawa *et al.* 1986). Silverback gorillas systematically protect group members from perceived threats whereas chimpanzee males do not (Tutin & Fernandez 1990). Female gorillas have a greater reproductive potential than chimpanzees (earlier age at first birth and shorter inter-birth interval), and this may be a result of living in a closed group (Tutin 1990). Perhaps the social adjustment of chimpanzees to food availability has disadvantages, if foraging alone when fruit is scarce either increases the risk of predation or increases the costs of maternal investment as male protection is not available.

In a complex frugivorous community, competition for preferred food occurs not only within, but also between, species. Of eight species of diurnal primate at Lopé, seven are frugivorous. Clear examples of competition for food are few, but the outcome appears to be determined by relative numbers rather than body size. Thus, although data are scarce, it seems possible that being a member of a group may provide an advantage in inter-specific competition for fruit sources, especially during periods of overall fruit scarcity.

The highly frugivorous diet of gorillas at Lopé shows the genus *Gorilla* has dietary flexibility, but even when fruit is abundant they continue to eat some foliage. Whether this is related to maintaining their complex gut flora and fauna, larger body size, a result of group living, or a combination of these factors, it is a feature that appears to differentiate them from chimpanzees. Detailed behavioural data on the gorillas and chimpanzees at Lopé are needed to quantify levels of intra- and inter-specific competition for food, and to assess the reproductive success of the two species. Continued study of the sympatric apes at Lopé should also shed light on the advantages and disadvantages of the different foraging strategies that were open to our hominoid and hominid ancestors.

We thank the L. S. B. Leakey Foundation, the World Wide Fund for Nature, the National Geographic Society, and especially the Centre International de Recherches Médicales de Franceville for supporting this research. Alphonse Mackanga and Joseph Maroga–Mbina assisted at Lopé. We are very grateful to the following colleagues who contributed to data collection: Catherine Bouchain, Jean-Yves Collet, Alick Cruickshank, Anna Feistner, Stephanie Hall, Rebecca Ham,

Fiona Maisels, Richard Parnell, Ann Pierce, Ben Voysey, Lee White and Dorothea Wrogemann; and to Lee White for constructive comments on the manuscript.

REFERENCES

Alexander, R. 1974 The evolution of social behavior. *A. Rev. Ecol. Syst.* **5**, 325–383.

Ashford, R. W., Reid, G. D. F. & Butynski, T. M. 1990 The intestinal faunas of man and mountain gorillas in a shared habitat. *Ann. trop. Med. Parasitol.* **84**, 337–340.

Chivers, D. J. & Hladik, C. M. 1984 Diet and morphology in primates. In *Food acquisition and processing in primates* (ed. D. J. Chivers, B. A. Wood & A. Bilsborough), pp. 213–230. London: Plenum Press.

Clutton-Brock, T. H. & Harvey, P. H. 1980 Primates, brains and ecology. *J. Zool.* **190**, 309–323.

Collet, J. Y., Bourreau, E., Cooper, R. W., Tutin, C. E. G. & Fernandez, M. 1984 Experimental demonstration of cellulose digestion by *Troglodytella gorillae*, an intestinal ciliate of lowland gorillas. *Int. J. Primatol.* **5**, 328.

File, S. K., McGrew, W. C. & Tutin, C. E. G. 1976 The intestinal parasites of a community of feral chimpanzees, *Pan troglodytes schweinfurthii*. *J. Parasit.* **62**, 259–261.

Freeland, W. J. & Janzen, D. H. 1974 Strategies of herbivory by mammals: the role of plant secondary compounds. *Am. Nat.* **108**, 269–289.

Gaulin, S. J. C. & Konner, M. 1977 On the natural diet of primates including humans. In *Nutrition and the brain*, vol. 1 (ed. R. J. Wurtman & J. J. Wurtman), pp. 1–86. New York: Raven Press.

Gautier-Hion, A. & Gautier, J. P. 1979 Niche écologique et diversité des éspeces sympatriques dans le genre *Cercopithecus*. *Terre Vie, Rev. Écol.* **33**, 493–508.

Goussard, B., Collet, J.-Y., Garin, Y., Tutin, C. E. G.& Fernandez, M. 1983. The intestinal entodiniomorph ciliates in wild lowland gorillas (*Gorilla gorilla gorilla*) in Gabon, West Africa. *J. med. Primatol.* **12**, 239–249.

Groves, C. P. 1970 Population systematics of the gorilla. *J. Zool.* **161**, 287–300.

Harvey, P. H. & Clutton-Brock, T. H. 1985 Life history variation in primates. *Evolution* **39**, 559–581.

Hiraiwa-Hasegawa, M., Byrne, R. W., Takasaki, H. & Byrne, J. M. E. 1986 Aggression toward large carnivores by wild chimpanzees of Mahale Mountains National Park, Tanzania. *Folia primatol.* **47**, 8–13.

Hladik, C. M. 1973 Alimentation et activité d'un groupe de chimpanzés réintroduits en forêt Gabonaise. *Terre Vie* **27**, 343–413.

Imai, S., Ikeda, S., Collet, J. Y. & Bonhomme, A. 1991 Entodiniomorph ciliates from the wild lowland gorilla with description of a new genus and three new species. *Eur. J. Prositol.* **26**, 270–278.

Milton, K. 1981 Distribution patterns of tropical plant foods as an evolutionary stimulus to primate mental development. *Am. Anthropol.* **83**, 534–548.

Milton, K. 1984 The role of food-processing factors in primate food choice. In *Adaptations for foraging in non-human primates* (ed. P. S. Rodman & J. G. H. Cant), pp. 249–279. New York: Columbia University Press.

Nishida, T. 1987 Local traditions and cultural transmission. In *Primate societies* (ed. B. B. Smuts, D. L. Cheney, R. M. Seyfarth, R. W. Wrangham & T. T. Struhsaker), pp. 462–474. University of Chicago Press.

Rogers, M. E., Maisels, F., Williamson, E. A., Fernandez, M. & Tutin, C. E. G. 1990 Gorilla diet in the Lopé Reserve, Gabon: a nutritional analysis. *Oecologia* **84**, 326–339.

Sarich, V. M. 1983 Appendix: retrospective on hominoid macromolecular systematics. In *New interpretations of ape and human ancestry* (ed. R. Ciocchon & R. S. Corruccini), pp. 137–150. New York: Plenum Press.

Shea, B. T. 1983 Size and diet in the evolution of African ape craniodental form. *Folia primatol.* **40**, 32–68.

Sussman, R. W. 1987 Species-specific dietary patterns in primates and human dietary adaptations. In *The evolution of human behavior: primate models* (ed. W. G. Kinzey), pp. 151–179. Albany: State University of New York Press.

Terborgh, J. 1983 *Five New World primates.* Princeton University Press.

Terborgh, J. 1986 Keystone plant resources in the tropical forest. In *Conservation biology: the science of scarcity and diversity* (ed. M. E. Soule), pp. 330–344. Sunderland, Massachusetts: Sinauer Associates.

Tsukahara, T. & Nishida, T. 1990 The first evidence for predation by lions on wild chimpanzees. In *13th Congress of IPS, abstracts*, p. 82. Nagoya.

Tutin, C. E. G. 1990 Social and biological parameters of first reproduction in African great apes. In *13th Congress of IPS, abstracts*, p. 156. Nagoya.

Tutin, C. E. G. & Fernandez, M. 1985 Foods consumed by sympatric populations of *Gorilla gorilla gorilla* and *Pan troglodytes troglodytes* in Gabon: some preliminary data. *Int. J. Primatol.* **6**, 27–43.

Tutin, C. E. G. & Fernandez, M. 1990 Responses of wild chimpanzees and gorillas to the arrival of primatologists: behaviour observed during habituation. In *Primate responses to environmental change* (ed. H. O. Box), pp. 187–197. London: Chapman and Hall.

Tutin, C. E. G., McGrew, W. C. & Baldwin, P. J. 1981 Responses of wild chimpanzees to potential predators. In *Primate behavior and sociobiology* (ed. A. B. Chiarelli & R. S. Corruccini), pp. 136–141. Berlin: Springer Verlag.

Williamson, E. A., Tutin, C. E. G., Rogers, M. E. & Fernandez, M. 1990 Composition of the diet of lowland gorillas at Lopé in Gabon. *Am. J. Primatol.* **21**, 265–277.

Wrangham, R. W. 1977 Feeding behaviour of chimpanzees in Gombe National Park, Tanzania. In *Primate ecology* (ed. T. H. Clutton-Brock), pp. 503–538. London: Academic Press.

Wrangham, R. W. 1979 On the evolution of ape social systems. *Social Sci. Inf.* **18**, 335–368.

Wrangham, R. W. 1987 Evolution of social structure. In *Primate societies* (ed. B. B. Smuts, D. L. Cheney, R. M. Seyfarth, R. W. Wrangham & T. T. Struhsaker), pp. 282–298. University of Chicago Press.

Wrangham, R. W. & Smuts, B. B. 1980 Sex differences in the behavioural ecology of chimpanzees in the Gombe National Park, Tanzania. *J. Reprod. Fert. Suppl.* **28**, 13–31.

Discussion

S. A. Altmann (*Department of Ecology and Evolution, University of Chicago, Illinois, U.S.A.*). The fact that the gorillas in this forest do not eat the fruit of oil palms is puzzling, as these fruits are nutritious and are eaten by chimpanzees. The oil from the fleshy mesocarp contains high levels of beta carotene and other carotenoid compounds. The composition of the fruits is very variable, with a range of about 37300–128700 μg beta carotene or its equivalent per 100 g oil (Leung 1968). If provitamin A is very abundant in the gorilla's diet from other sources, or if these primates are particularly sensitive to carotenoids, their intake may border on hypervitaminosis and thereby condition them against oil palm fruit.

Reference

Leung, W. W. 1968 *Food composition table for use in Africa.* (306 pages.) FAO and USDHEW.

D. A. T. Southgate (*Institute of Food Research, Norwich, U.K.*). Professor Altmann raised the issue of possible effects of excessive carotene intakes. In humans excessive intake does not produce vitamin A toxic effects: the subject merely looks jaundiced from the high circulating levels of carotenoids.

The avoidance of fatty foods by gorillas creates a need to consume a very large bulk of food and appears a rather unusual foraging strategy.

M. E. Rogers. When the authors talk about avoidance of fatty fruits, they are referring to very high fat content of 20–76% dry mass. When it is suggested that gorillas eat some fatty foods in other studies (Calvert 1985), much lower fat levels were being referred to.

Reference

Calvert, J. J. 1985 Food selection by western gorillas (*G. g. gorilla*) in relation to food chemistry. *Oecologia* **65**, 236–246.

I. Crowe (*23 Lockhart Close, Dunstable, Bedfordshire, U.K.*). Why do gorillas continue to eat leaves when other favoured foods are available?

M. E. Rogers. Leaves are high-protein foods, and this must be important as both fruit and pith are low in protein. Also, gorillas choose leaf foods from common herbaceous plants that have low levels of digestion inhibitors (total phenols and tannins). It may be desirable from the point of view of digestion to mix high- and low-tannin foods at all times.

I. Crowe. Dr Foley has suggested that climatic changes in Africa (i.e. long dry seasons) affected early hominid diets. I suggest a similar factor might explain the extinction of *Australopithecus boisei* and similar sub-species, given that *A. boisei* was probably a folivore subject to the same dietary constraints: unable to risk diversifying too much, being dependent upon maintaining a high leaf content in its diet, to avoid destroying the gut flora and fauna.

C. E. G. Tutin. The parallel between sympatric hominids and the sympatric apes at Lopé is interesting. Our data show that gorillas at Lopé can subsist on a predominantly folivorous diet, and are dependent on an abundant supply of leaf foods during the annual dry season. Sufficient leaf foods are available throughout the year in equatorial regions and it seems likely that this resource defines the geographical range of the large-bodied *Gorilla*. Chimpanzees can subsist in habitats with a prolonged dry season which are dominated by deciduous vegetation.

A. Whiten (*Scottish Primate Research Group, University of St Andrews, U.K.*). The authors have undertaken very extensive analyses of the chemical composition of gorilla (and chimpanzee?) foods. Do these allow the authors to make comparisons with norms of composition in the diets of different populations of baboons such as Whiten *et al.* (this symposium) show in table 5, or is such comparison defeated by the observational constraints on recording intake?

M. E. Rogers. Observational constraints have certainly prevented the ranking of food items by proportion of total intake, or of feeding time; but they can be ranked by frequency of occurrence on trails and in faeces. By using these data we can provide a mean diet composition for Lopé gorillas. We do not have comparable data for chimpanzees at Lopé.

S. A. Altmann. To what extent are the primates the authors have studied responsible for dispersing the seeds of the plants on which they feed? The faeces of Amboseli baboons contain intact seeds of *Acacia tortilis* and *Balanites aegyptica*, sometimes deposited, complete with fertilizer, at considerable distances from the source plant.

C. E. G. Tutin. Both gorillas and chimpanzees at Lopé disperse the intact seeds of most of their fruit foods. Gorillas disperse seeds of at least 65 species (Tutin *et al.* 1991). The large gut size of both species of ape and their wide ranging makes them seed dispersers 'par excellence', and they undoubtedly play an important role in the dynamics and regeneration patterns of their tropical forrest habitats.

Reference

Tutin, C. E. G., Williamson, E. A., Rogers, M. E. & Fernandez, M. 1991 A case study of a plant–animal relationship: *Cola lizae* and lowland gorillas in the Lopé Reserve, Gabon. *J. trop. Ecol.* **7**, 181–199.

N. L. Conklin (*Department of Anthropology, Harvard University, Cambridge, Massachusetts, U.S.A.*). Chimps are definitely seed dispersers. We are hoping to study the details of seed dispersion by chimps in Kibale in the near future.

L. Barrett (*Department of Anthropology, University College London, U.K.*). There have been several reports, notably by Richard Wrangham, which suggest that chimpanzees at Gombe seek out the leaves of certain plant species, e.g. *Aspilia* spp., specifically for their pharmacological effects, and it is known that, although the baboons at Amboseli avoid the fruits of *Solanum incanum* (which contain high levels of the alkaloid, solanin), those at Laikipia regularly eat the unripe fruits of this species, possibly for 'medical' purposes. Has this phenomenon been noted among the apes at Kibale and Lopé? More generally, how prevalent or important is the selection of plant species for pharmacological, rather than nutritional, purposes among monkeys and apes?

M. E. Rogers. Gorillas eat a lot of ginger (African ginger, *Aframomum* spp.). Commercial ginger (*Zingiber officinale*), which is an Asiatic species, has antihelminthic properties so, if *Aframomum* does also, then gorillas may be regularly consuming a plant with medicinal properties. *Aframomum* certainly contains terpenes. I am not suggesting that they eat it selectively for medicinal reasons, just that they eat so much of it anyway, it may well have therapeutic effects in terms of their parasite load.

N. L. Conklin. The phenomenon is currently being studied in Kibale, but it is too early to say more than that. The topic is reviewed by Goodall & Wrangham (1989).

Reference

Goodall, J. & Wrangham, R. 1989 In *Understanding chimpanzees* (ed. P. G. Heltne & L. A. Marquadt). Cambridge, Massachusetts: Harvard University Press.

Dietary and foraging strategies of baboons

A. WHITEN[1,2], R. W. BYRNE[1], R. A. BARTON[1,2]†, P. G. WATERMAN[1,3] AND S. P. HENZI[1]‡

[1] *Scottish Primate Research Group, Department of Psychology, University of St Andrews, St Andrews, Fife KY16 9JU, U.K.*
[2] *Institute of Primate Research, P.O. Box 24481, Nairobi, Kenya*
[3] *Phytochemistry Research Laboratory, University of Strathclyde, Strathclyde G1 1XW*

SUMMARY

As large-bodied savannah primates, baboons have long been of special interest to students of human evolution: many different populations have been studied and dietary comparisons among them are becoming possible. Baboons' foraging strategies can be shown to combine high degrees of flexibility and breadth with selectivity. In this paper we develop and test multivariate models of the basis of diet selection for populations of montane and savannah baboons. Food selection is positively related to protein and lipid content and negatively to fibre, phenolics and alkaloids. Seasonal changes in dietary criteria predicted by these rules are tested and confirmed. Although nutritional bottlenecks occur at intervals, a comparison between long-term nutrient intakes in four different populations indicates convergence on lower degrees of variation than exist in superficial foodstuff profiles.

1. INTRODUCTION

Partly because of the relevance of the savannah primate niche to human evolution (Peters & O'Brien 1981) and partly for other reasons such as the superior observation conditions this habitat offers, baboon research has been pursued during the last thirty years at numerous locations across Africa (Dunbar 1988). Many studies have quantified the profile of foodstuffs consumed (table 1), so that in baboons more than any other non-human primate, we can start to appreciate the true range of dietary variation.

In recent years primatologists have become more ambitious, attempting to quantify diet not only at the level of foodstuff profile, but in terms of actual nutrient ingestion. However, quite apart from the extensive laboratory analyses this involves, it is an exacting task to follow a focal individual in the field and make a continuous record of the number, size and identity of each bite of food (which can be one per second), even when facilitated by the use of hand-held micro-computers (Whiten & Barton 1988; Byrne *et al.* 1990*b*). Accordingly these 'fine-grain' analyses are still rare in primate research. Barton (1992) found just eight which provided accurate measures of daily dry mass intake.

However, in baboon research we are again uniquely fortunate in that results of such work are becoming available from parallel efforts at five different sites in east and southern Africa. In the main part of this paper we present results from this new work, including our own conducted at two of these sites: one in savannah, the other in a more marginal montane environment.

2. THE FORAGING STRATEGY OF BABOONS: AN OVERVIEW

Characterization of baboons' foraging strategy as adaptably broad, yet selective, may sound paradoxical: yet it is probably the combination that permits a primate like this to succeed in the habitats concerned (Whiten *et al.* 1987; Norton *et al.* 1987). In what follows we summarize the principal dimensions of baboons' approach to foraging.

(a) Adaptability

Although baboons' modal habitat can be fairly characterized as the band of savannah that sweeps from west, through east, to southern Africa, baboons also exhibit successful invasions of a variety of types of habitat on its margins, including desert, swamp, forest and montane environments. Foraging profiles described even at the level of very gross food types show remarkable variations across these habitats: for the most common food types the ranges are 3–74 % (fruit), 1–53 % (subterranean items) and 8–53 % (leaves) (table 1). At this level of analysis, baboons' foraging strategy can clearly be characterized as highly flexible and adaptable.

(b) Dietary diversity

The ability to adapt to varying habitats may be a product of flexibility within the foraging strategy

† Present address: Department of Psychology, University of Sheffield, Sheffield S10 2TN.
‡ Present address: Department of Psychology, University of Natal, Durban 4001, R. South Africa.

Table 1. *Local variations in the foraging profiles of baboons*

(Sites are ranked by % fruit in the diet (the most common food type overall).)
1. * Mt Assirik, Senegal (*P. papio*), Sharman (1981); 2. * Gombe, Tanzania (*P. anubis*), J. Oliver; 3. * Suikerbosrand, S. Africa (*P. ursinus*), C. Anderson; 4. Mikumi, Tanzania (*P. cynocephalus*), Norton *et al.* (1987); 5. * Cape Reserve, S. Africa (*P. ursinus*), Davidge (1978); 6. * Bole, Ethiopia (*P. anubis*), Dunbar & Dunbar (1974); 7. * Amboseli, Kenya (*P. anubis*), Post (1978); 8. Laikipia, Kenya (*P. anubis*), Barton (1989); 9. * Ruaha, Tanzania (*P. cynocephalus*), Ramussen (1978); 10. * Gilgil, Kenya (*P. anubis*), Harding (1976); 11. * Drakensberg, S. Africa (*P. ursinus*), Whiten *et al.* (1987). * Data from Dunbar (1988), p. 295.)

	percentage of time feeding on item at different study sites (1–11)													
food type	1	2	3	4	5	6	7	8	9	10	11	range (%)	mean (%)	coefficient (%) of variation
fruits	74	49	43	43	42	41	27	23	16	10	3	3–74	34	61
underground items	3	7	39	12	16	1	33	15	52	27	53	1–53	23	79
leaves	9	14	8	14	25	41	15	27	19	53	26	8–53	23	61
flowers	9	2	7	20	12	12	5	21	1	3	14	1–20	10	72
animals	1	13	3	?	3	4	1	1	9	2	4	1–13	4	103

associated with the modal savannah habitat itself, where baboons exploit an unusual diversity of foods. The most straightforward measure of dietary diversity is the number of different types of food item eaten. Norton *et al.* (1987) expanded their baboon food list in each of five years of study to include 1–6 parts of 185 species of plants.

Another aspect of diversity may be even more important in the maintenance of baboons' particular feeding niche: the ability to extract nutrients from almost all compartments of the environment. At both of our study sites baboons spend the major part of their foraging time ranging through and selecting foods from the seasonally productive herb and shrub layer: but additionally, feeding niche separation from ungulate competitors in this layer is achieved by the capacity to probe the environment in other ways. This includes arboreal foraging which extends the range of leaves, flowers and fruit taken, and also makes available special foods like exudate and birds' eggs; and subterranean foraging, which involves digging up the fleshy bases, roots and storage organs of a variety of herbs, grasses and sedges. Exploitation of the full diversity of these food sources means that baboons can cope with the severe seasonality that typically dries and kills the herb layer for months at a time in these habitats, forcing ungulates into substantial migrations that the baboons can avoid. At our savannah site the arboreal compartment (particularly *Acacia* flowers and seeds) plays an important role (42 % of annual intake by mass), and underground items a more minor one (5.6 % by mass, although 15 % of feeding time; Barton 1989): at our high altitude montane site the reverse is the case and subterranean items account for virtually all the diet in the cold, dry, winter season (Whiten *et al.* 1987).

In a survey of diets in 120 primate species, Harding (1981) found that the typical savannah species, baboons and vervet monkeys, had the most diverse diets (utilizing all nine of the compartments he assessed), followed by woodland – savannah chimpanzees. However, at two sites where baboons and chimpanzees are sympatric, baboons still exhibit the greater diversity of food profile (Peters & O'Brien 1981, 1982).

(*c*) ***Selectivity***

Although earlier studies tended to equate baboons' breadth of diet with lack of discrimination, considerable selection can be demonstrated, at a number of levels.

First, baboons utilize only a proportion of the species available. Norton *et al.* (1987) estimated that the 185 species of plant used by Mikumi baboons are selected from among approximately 700. Selection may be expressed at a fine level of discrimination, between species of the same genus which to a human eye appear superficially very similar. Thus, for example, Drakensberg baboons select the corms of just one of three similar species of *Hypoxis* (table 2).

At another level, it is often only one or a few parts of a food species that are consumed. Separating this may

Table 2. *Selection among corms of related species of Hypoxis*

(All data are percentage of dry mass except water (which is percentage of net mass), starch and alkaloid (both of which are graded on a 3-point scale). Phenol. = total phenolics, tann. = condensed tannins, alk. = alkaloids.)

species	water	protein	lipid	starch	phenol.	tann.	alk.	fibre
H. gerrardii[a]	78.2	3.23	7.0	'3'	0.78	0	0	11.3
H. obtusa	68.6	1.42	6.0	'1'	1.46	0	0	9.8
H. rigidula	70.0	2.46	1.0	'1'	11.26	0	0	12.0

[a] Species consumed.

Table 3. *Composition of selectively processed food parts (Acacia)*

		water	protein	lipid	phen.	tann.	alk.	fibre
A. seyal	seed kernel[a]	74.0	49.2	7.5	0.5	0.4	0	4.3
	seed skin	66.7	15.1	3.0	7.5	12.1	0	22.7
	pod	68.2	14.4	5.0	5.6	2.3	0	37.9
A. nilotica	seed kernel[a]	74.5	47.0	11.5	0.6	0.8	0	3.1
	seed skin	59.0	12.0	4.0	7.8	18.0	0	21.8
	pod	—	11.6	12.0	7.2	7.6	'2'	11.5

[a] Denotes parts consumed. Other conventions as for table 2.

be relatively straightforward for a primate like a baboon, as in plucking a flower or fruit. In a number of other cases both hands are used, sometimes in conjunction with the teeth, to process the food, peeling and discarding certain components. For example, in three of six species of *Acacia* used as foods by Laikipia baboons, pods are torn from the tree and, using both hands and teeth together, sliced open and the seeds nibbled out: within the mouth the skins of the seeds are separated and pushed out using the tongue. By such processing the baboons exert considerable control over the composition of the material they finally ingest (table 3), in a fashion simply not open to less dextrous competitors.

3. IDENTIFYING THE BASIS OF DIET SELECTION

The contrasts in composition summarized in tables 2 and 3 suggest some of the reasons why baboons select the species and parts they do. We now tackle the thorny question of how more systematically and comprehensively to specify whatever rules underlie the dietary 'decisions' being made.

(*a*) *Background and rationale*

One of the first nutritional comparisons between selected and discarded food parts in primates was offered by Hamilton *et al.* (1978) for three food types in chacma baboons. Parts chosen for ingestion contained relatively high proportions of protein (also lipid in two cases) and low proportions of fibre (as in table 2).

Aside from this pioneering study, it was primatologists studying arboreal folivores who pursued food versus non-food comparisons extensively (see Milton 1980). The general technique has been the simple one we have already described: foods are compared with non-food control items which appear superficially similar and which present an obvious puzzle as to why they are avoided. A number of constituents may be analysed and, for each one, a food versus non-food difference is tested for statistical significance. Common components examined include potential macronutrients, fibre and secondary compounds. The latter include digestion inhibitors (notably condensed tannins) and toxins such as tannins and alkaloids, which are known to act as anti-herbivory defences in certain contexts (Waterman 1984).

These comparisons show recurring trends, particularly for lower levels of protein and higher levels of fibre in non-foods, implicating these constituents as among the determinants of choice. However, such component-by-component comparisons must ultimately be of limited usefulness. In species like baboons where the diet is so diverse, there is enormous overlap between the proportions of any one component (even protein) when we compare foods with a variety of non-foods. To complicate matters further, several other components differ markedly in some specific food versus non-food contrasts, suggesting a role in these choices at least.

All this suggests what is in any case plausible from first principles: that the basis on which certain items are chosen or even preferred as foods constitutes an equation in which a certain balance of elements is aimed at. Thus, for example, a relatively high proportion of a single nutrient might make an item as attractive as another with lesser proportions of two different nutrients; or an item with a relatively low proportion of nutrient may be acceptable if it has little in the way of negative components like toxins; or given two items with similar nutrient and toxin profiles, it may then be the level of fibre which finally distinguishes one as a food.

We have used discriminant analysis to derive just such multivariate equations expressing the rules of diet selection.

(*b*) *Methods*

Two populations of baboons have been studied. Since 1986 a group of approximately 100 olive baboons

(*Papio anubis*) has been studied in savannah habitat on the Laikipia plateau, Kenya. Chacma baboons (*P. ursinus*) studied earlier in the Drakensberg mountains on the S. African–Lesotho border sustained relatively low population densities and formed smaller groups: one of approximately nine and another of 14 were studied over an 18-month period.

At each location a variety of standard methods have been used to gather data on behaviour, demography, food availability and nutrient composition (Barton 1989; Barton *et al.* 1992*a*, *b*; Byrne *et al.* 1990*a*, *b*; Whiten *et al.* 1987, 1990).

The principal analyses to be discussed below involve contrasts between plant food and non-food controls. Non-foods included the two types distinguished in the previous section: (i) parts of food plants discarded or avoided; and (ii) parts of species that are not eaten, corresponding to parts of species that are eaten. In the latter case we chose controls that appeared most similar to the item eaten, as in the example already illustrated in table 2. The assay techniques for these samples are described in Whiten *et al.* 1990 and in more detail in Choo (1981). In summary, they included water (air drying at 37 °C to constant mass), total nitrogen (micro-Kjeldahl), lipid (ether extraction), starch (4-point scale based on iodine stain), fibre (acid-detergent), condensed tannins (as quebracho-tannin equivalents, through modified butanol–HCl method), total phenolics (tannic acid equivalents, through Folin-Denis method) and alkaloids (3-point scale utilizing reactions with Dragendorf's reagent). Protein was calculated at $N \times 6.5$, and total carbohydrate was then calculated by subtraction.

Such data from the Drakensberg site have been subjected to a number of stepwise discriminant analyses (SPSS 1983) to establish what components are critical, in combination, for discriminating food from non-food items. For each item, the discriminant procedure computes a function:

$$\text{Food value score} = C + A \cdot x + B \cdot y + D \cdot z \ldots$$

in which C is a constant, A, B are proportional amounts of components (e.g. protein) and x, y are coefficients computed so as to maximize the difference between mean food value scores for foods and non-foods. Coefficients can be positive (valued components) or negative (avoided components).

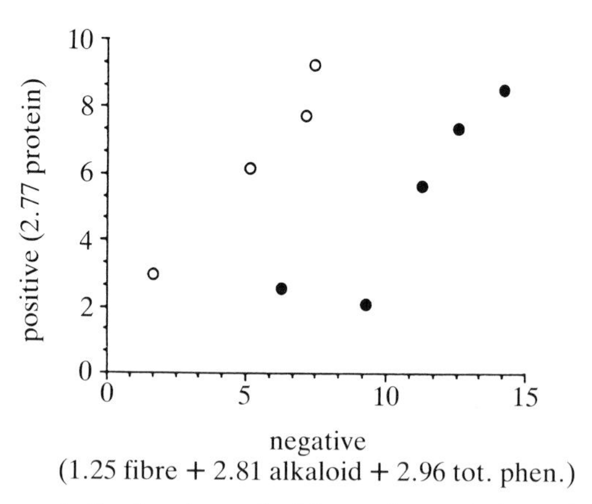

Figure 1. Food and non-food leaves (open circles and closed circles, respectively) plotted against axes separately representing the positively and negatively weighted components contributing to each item's food value score. Coefficients (weightings) are standardized. Note that neither protein alone, nor the negative complex, can differentiate foods from non-foods: only their combined effect achieves this.

(*c*) *Selection rules for major food types*

For four major food items separate comparisons of foods with control non-foods have been done. The end result can be appreciated in an example (figure 1). The sum of the positively-weighted elements in the discriminant function are plotted against those negatively weighted, producing maximal separation between food and non-food leaves. Table 4 shows the weights (coefficients) computed separately for leaves, corms, inflorescences and pale fleshy bases of leaves and stems.

The coefficients indicate that in the selection of different types of foods, different rules apply: only in the case of leaf bases is lipid a significant factor, for example. Just one factor, protein, was important in all cases, although for bases it was outweighed by the coefficients for lipid and fibre content. A number of components consistently failed to contribute to discrimination rules: water content, starch, condensed tannins and total carbohydrate.

(*d*) *Seasonal change in selection rules*

An alternative way of representing graphically such rules is to show in histogram form the relative weights

Table 4. *Standardized discriminant function coefficients distinguishing food types from non-food matched control items*

(Presentation of standardized coefficients here permits direct comparison of the 'importance' of any component within, although not between, analyses. Thus variance in protein, total phenolics and fibre are of roughly equal weight in making discriminations for both leaves and inflorescences.)

	leaves	corms	bases	inflorescences
protein	2.8	2.23	0.11	0.67
lipid	—	—	1.12	—
phenolics	−3.0	—	—	—
alkaloid	−2.8	—	—	−0.79
fibre	−1.3	—	−1.2	−0.99
discrimination of foods from non-foods (%)	100	100	100	100
n	9	10	11	10

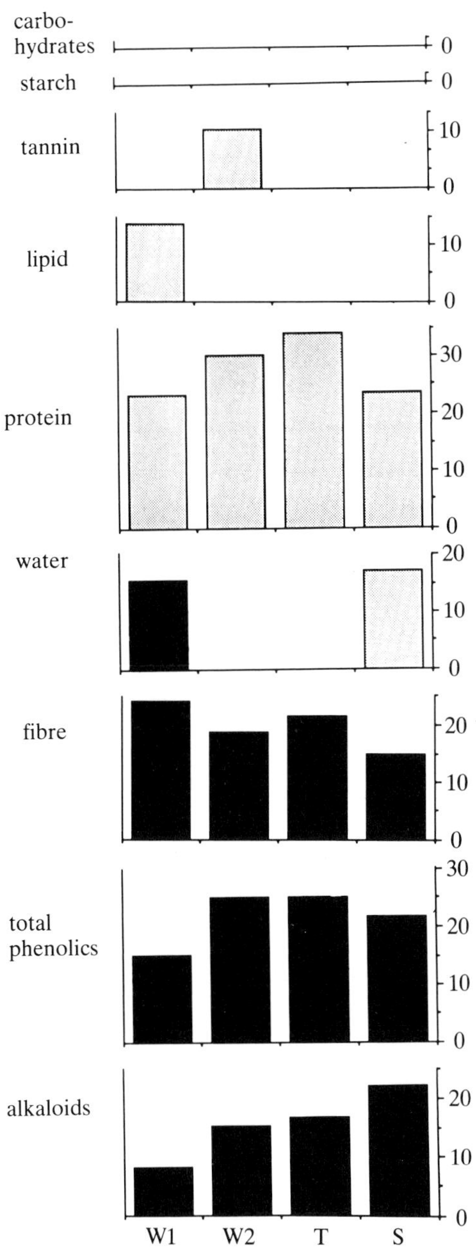

Figure 2. Size of standardized discriminant coefficients, expressed as percentages to show their relative importance in each of four seasons. Because of its inconsistent role, water was dropped from prediction-across-season analyses shown in subsequent figures. Open bars, positive weight; filled bars, negative weight.

attaching to different components. Figure 2 does so for four consecutive periods from winter to spring, during which a flush of fresh leaves and flowers gradually develops (Byrne *et al.* 1990*b*). For each season a separate function has been computed, combining all foods eaten in that season and contrasting them with a sample of appropriate non-food controls. We might expect this more global rule to combine those components significant in the case of the separate comparisons shown in table 3, and indeed this is the case. Of these, four components – protein, fibre, total phenolics and alkaloids – remain significant in all seasons, whereas lipid is picked out only in the winter period, consistent with the fact that underground storage organs are particularly utilized at that time. By

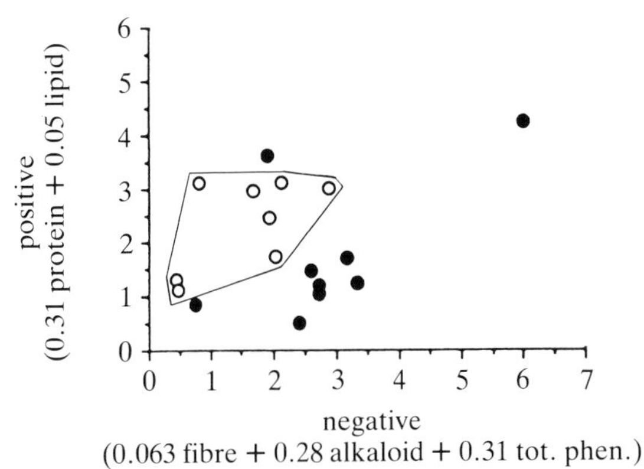

Figure 3. Winter foods (greater than 1 % of diet, open circles) and non-foods (closed circles) plotted on axes showing unstandardized coefficients. The one non-discriminated non-food is an unripe fruit, presumably distinguished from the ripe, eaten fruit by an element we have not yet measured. This item is therefore dropped from analyses shown in subsequent figures.

contrast, (avoidance of) alkaloids receives progressively heavier weightings as inflorescences and leaves protected in this way become increasingly abundant. This is despite the fact that 'alkaloids' were estimated on only a crude rating scale which did not discriminate the many different specific toxins (and toxicity levels) potentially involved.

(*e*) *Testing predictions of the rules*

The status of the 'rules' as we have discussed them so far is essentially descriptive. They simply represent the 'best' overall numerical specification of what distinguishes any particular set of foods and non-foods.

However, we might elevate the rules to the status of a model of the preferences underlying baboons' behaviour if further predictions can be generated and tested. Two ways in which this can be done rely upon the fact that the discriminant procedure works by computing the relative 'attractiveness' of each food and non-food: the food value score. One test of the accuracy of the computed rules is thus that foods with high food value scores should be particularly valued by the baboons, and if monopolizable, they should be the subject of the most intense feeding competition. This test has not yet been applied, although it is apparent that one prediction from these rules derived from plant foods, that vertebrate meat (with high values of protein and lipid and little or no fibre, phenolics or alkaloids) should be highly valued when available, is clearly confirmed (Strum 1981).

Another prediction rests on optimal foraging theory: as food value scores improve seasonally, the threshold of these scores at which the food versus non-food cut-off is made by the animals should rise. To test this, we first show (in figure 3) the general rule describing winter food choice for all analysed foods contributing more than 1 % of feeding records (accounting for 90.3 % of feeding). The polygon describing these points is then preserved in figure 4 and overlaid with the scores for spring foods (each greater than 1 % of the diet and together accounting for 69.9 % of feeding) and non-

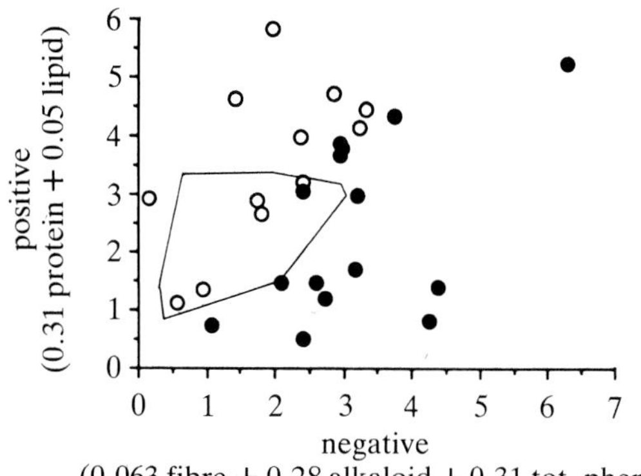

Figure 4. Spring foods (open circles) and non-foods (closed circles; greater than 1 % of diet) superimposed upon polygon for winter foods taken from figure 3. Note the trend in both foods and non-foods to upper left ('attractive') quadrant.

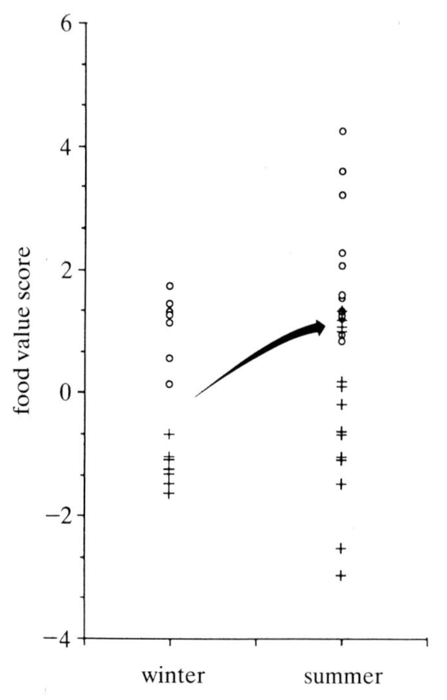

Figure 5. Seasonal change in distribution of food value scores for foods (circles) and non-foods (crosses), using winter coefficients for both seasons. Change in the food versus non-food 'cutoff' is indicated by the arrow.

foods. Both food and non-food scores rise higher in the direction of positive weightings, and the threshold between them appears to rise also, as predicted. This change in cut-off is shown more clearly in a single-dimension plot (figure 5). To check that this effect is robust, Figure 6 repeats the contrast using the coefficients computed from spring data. In this case there is naturally a clearer separation in the spring, (and applying these coefficients to the winter foods and non-foods conversely produces a less clear separation than in figure 5). This result offers strong support for the model because the two low-value foods that drop out of the foraging profile in spring are far from insignificant for the animals in mid-winter: one, a grass base (*Sporobulus congoensis*), accounted for 49.3 % of the foraging records for one group; the other, a sedge base

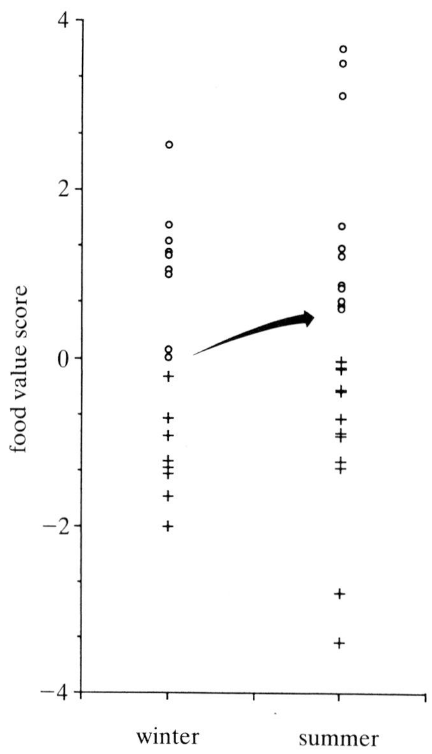

Figure 6. As for figure 5 but using summer coefficients for both seasons.

(*Scirpus falsus*), accounted for 73.2 % of the foraging records of the other group, studied at a high-altitude site.

(*f*) *Dietary correlates of selection ratios in Laikipia baboons*

Although the selection rules described above are derived from only food versus non-food contrasts, the predictions confirmed in the previous section suggests that they may have relevance to preferences among the foods eaten. To test this properly requires the computation for each food of a selection ratio, correcting the proportion of a food in the diet by the availability of it in the environment (at Laikipia the two are correlated, $r = 0.53$, $p = 0.012$). Only for the Laikipia site can we do this well. Discriminant analyses like those outlined above for Drakensberg baboons have not yet been done at Laikipia. We therefore rely on the finding that regular features of the Drakensberg selection rules include positive weightings for protein and negative weightings for fibre of a similar order of magnitude, and have therefore computed protein:fibre ratio that *is* available for Laikipia. Using figures for ten major categories of diet, it was confirmed that selection ratio is well predicted by protein:fibre ratio ($r = 0.829$, $p < 0.001$; Barton 1989).

4. DIETS CONSUMED: INTER-POPULATION COMPARISONS

(*a*) *Proportions of nutrients in the diet*

What dietary profiles result from the selection rules outlined above? Instructive, although as yet far from perfect, comparisons can now be made between four

Table 5. *Diet composition in four baboon populations*

(Data are mean [and median] percentage of dry mass except water, which is percentage of fresh mass. Laikipia means are for 13 most common foods (Barton *et al.* 1991), accounting for 68 % of feeding time. Drakensberg means are for eight most common foods, accounting for 68 % of feeding time. Amboseli means are for values given in Altmann *et al.* (1987), using eight most common foods, accounting for 68 % of feeding time, given by Post (1978, table 61). Mikumi means are for all plants tested (Johnson 1989, table 4.22). Ranges, rounded to whole numbers, are for all foods tested.)

	Laikipia	Mikumi	Drakensberg	Amboseli
water	70.9 (7–94) [78.7]	—	63.8 (34–93) [77.6]	57.6 [70.4]
protein	21.2 (5–50) [19.1]	13.68 (1–39)	8.44 (2–17) [7.3]	14.4 [15.8]
lipid	6.3 (2–14) [5.0]	5.25 (2–20)	6.3 (1–15) [5.0]	4.4 [4.1]
phenolics	1.7 (0–11) [1.3]	3.20 (0–23)	1.3 (0–13) [1.0]	—
tannin	1.4 (0–15) [0]	2.53 (0–49)	0.6 (0–2) [0]	—
fibre	20.9 (1–56) [24.0]	26.29 (5–55)	21.5 (1–40) [25.3]	22.1[a] [21.8]
n (for range)	66	33–192	39	—

[a] Different method used.

different sites (table 5). These include our Drakensberg and Laikipia sites, together with Amboseli (Kenya) and Mikumi (Tanzania), at both of which long-term studies of yellow baboons (*P. cynocephalus*) have been made. Phytochemical analyses from Amboseli are the most extensive (Altmann *et al.* 1987; Altmann 1990); however, the other three studies, despite covering a restricted range of constituents, offer the advantage of identical methods applied at Waterman's laboratory or directly derived from them.

There are several ways in which dietary comparisons might be attempted. The ideal would be to compute proportions of nutrients based on measures of absolute intake, but opportunities for comparisons on this basis are very restricted as yet (see §4*b*). The approach taken in table 5 is therefore the simpler one of examining the proportional composition of the most common food plants. At all four sites, two thirds of the diet is accounted for by a relatively small number of food types: just 8–13 for the three sites for which information is readily available. For these three sites, mean and median values for nutrient analyses for these items that constitute the bulk of the diet are given in table 5: for the fourth site, only means for the whole sample of foods tested are available. In three cases it is also possible to give the total range of variation.

These figures should therefore be interpreted as 'typical' nutritional profiles, together with the range over which they vary, for baboon foods at the sites concerned, rather than as exact specifications of the overall nutrient intake profile (see §4*b*). As such, they are offered as useful bases for future comparisons with data for other species. In their own right they show at least two important facts.

First, at all sites the range of variation of any one component – whether nutrients, fibre or secondary compounds – is very great. This is likely to be one of the reasons that diet selection rules have to be specified in multivariate terms, because ranges of values for any one component in non-foods can be shown to have a huge overlap with those in foods (Whiten *et al.* 1990). To give just one example, protein in Drakensberg non-foods ranged from 0.9 to 12.7 % compared with 2.1–16.5 % for foods.

Second, although some inter-population differences in the average figures are substantial, they show less variation than at the level of foodstuffs, where variability has already been emphasized. Despite a large difference between the mean protein scores of our Drakensberg and Laikipia baboons, the coefficient of variation across the four samples is 36 %, less than the coefficients of variation for basic foraging strategies shown in table 1. There is thus an indication that through these very variable foraging profiles baboons in different habitats converge on a more limited range of nutrient consumption.

(*b*) *Absolute nutrient intake*

Where data have been collected not only on the relative frequency of items in the diet, but on complete records of amounts ingested, it becomes possible to specify diet in absolute terms. To date, comparisons at this level can be made for protein intake at three of the baboon sites concerned.

This is not a straightforward task and the following must be regarded as a preliminary attempt. A number of differences in methodology are likely to introduce error. One is observational method, which was not identical in the three cases: however, the goal in all was to specify absolute intake and this is probably not the greatest difficulty for comparative purposes. Second, the data are published for focal animals of varying masses: here they are re-expressed in terms of an 'average adult' of 15 kg. Third and perhaps most problematic is the difficulty that in each case data are reported for part of a year, in habitats where we know seasonality can be severe.

Protein intake, in grams per day for a 15 kg baboon, is computed at 69.5 g for Amboseli (based on Stacey

(1986), table 3) and 88.2 g for Laikipia (based on Barton (1989), table 6.4). As these data are for periods wetter than the average, the fairest comparison in the Drakensberg data appears to be for the spring period (Byrne *et al.* 1990*b*), for which the figure is 75.7 g. The coefficient of variation across these figures for protein intake is then just 22 %, consistent with convergence on a more limited range of variation in nutrient intake than exists at the level of foodstuff profile or even at that of the composition of common foods (§4*a*). However, the conclusion that protein intake is consistent across populations in (wet) periods of relatively high availability must be contrasted with the likelihood of significant seasonal fluctuation. This becomes clear when we note that the corresponding protein intake figure for the severe winter and winter–spring transitional periods in the Drakensberg is only 28.2 g (Byrne *et al.* 1990*b*).

These figures are encouragingly consistent with the nutritional requirements in Oftedal's table 5 (p. 166). The winter Drakensberg intake is relatively low by this standard; the other, overall intakes reported above, are relatively high, but we must remember that the natural diets incorporate secondary compounds that are likely to reduce protein digestibility.

5. SUMMARY AND DISCUSSION

We have outlined a number of fundamental respects in which baboons are constrained by, and adapted to, their particular food niche. It seems likely that these would not have been avoided by early hominids in their first invasions of the habitat. Flexibility is observed in the readiness to adopt widely differing dietary profiles with respect both to the severe seasonal changes that characterize savannah, and to a variety of habitats marginal to savannah. Another dimension of dietary flexibility is the breadth of the food profile. Baboons' use of underground storage organs is an important tactic taken to further extremes by present-day savannah hunter-gatherers who can harvest a deeper layer by the use of a simple digging stick (Peters & O'Brien 1981). By contrast with the breadth of diet, precise visually guided reaching and manipulation permit a high degree of selectivity, starting with harvesting and continuing through manual and oral processing of the food.

The computation of rules of diet selection as discriminant functions that express the combined roles of constituents of foods in selection has been shown to be workable and to generate predictions which have been tested and supported. For Drakensberg baboons, protein, fibre, total phenolics and alkaloids were found to play almost equal roles in distinguishing foods from non-foods. The significance of protein and fibre is consistent with the food preferences of Laikipia baboons also, and with previous studies of primates exploiting the superficially very different niche of aboreal folivory (see, for example, Milton (1980); McKey *et al.* (1981)). However, the negative role of tannin found in the latter has been less emphasized in other studies of colobine monkey diet (Waterman *et al.* 1988; Davies *et al.* 1988) and was not found for mountain baboons, for whom phenolics and alkaloids were more important inhibitory influences. Phenolics and alkaloids may be subject to detoxification in the foregut of specialist colobine folivores, a facility probably absent in monogastric primates like baboons.

This analysis of diet selection rules must be regarded as a first approximation for a number of reasons. We have used only a small subset of the components which could in principle be examined, and even the categories we have used are not unitary ('alkaloids' is most obviously a heterogenous category). However, the inclusion of further components in future analyses can be expected essentially to refine an approach which is already justified by its empirical success in discriminating foods and non-foods by using the basic categories of component described here.

The approach draws an important distinction between the compositional basis for choice, and the dietary composition achieved. In the case of Drakensberg baboons, the latter includes significant quantities of starch and lipid, but starch does not figure in the selection rules and lipid does so in only a small way. Whiten *et al.* (1990) noted that this suggests that 'by following certain rules, baboons get, into the bargain as it were, a diet adequately provided with other components (such as starch) which in some foods are associated with the choice element'.

Such dietary 'rules of thumb' represent evolved adaptations to the nutritionally critical components in a certain ecological niche. In the case of baboons (and by extension in ancestral hominids), this niche is characterized by high levels of fibre and, at least in certain habitats and seasons, low levels of protein. It may be that the particular rules shown by our analyses are adaptations to an unusual degree of protein deprivation during the Drakensberg winter (table 5), although the Laikipia selection ratio correlations suggested similar priorities even where foods are richer in protein. Although in principle we might expect the negative weighting of fibre to be nonlinear (a certain amount providing optimal gut passage times), in a dietary environment producing a high input of fibre the rule that evolves might be less sophisticated, and open ended: 'maximize the protein:fibre ratio'. Such a rule should lead to consistent choice of minimal-fibre options if these became available as an evolutionary novelty. This may be what we see in baboons that gain access to food waste tips at African game lodges (and also perhaps in those descendants of savannah primates who frequent the restaurants in those lodges?)

We thank S. Altmann and F. Marsh for comments on previous versions of the manuscript. We are grateful for the support of the Science and Engineering Council, U.K. and the Institute of Primate Research, Kenya, and we thank Natal Parks Board and the Office of the President, Nairobi, for permissions to study baboon populations.

REFERENCES

Altmann, S. A. 1990 Diets of yearling female primates (*Papio cynocephalus*) predict lifetime fitness. *Proc. natn. Acad. Sci. U.S.A.* **88**, 420–423.

Altmann, S. A., Post, D. G. & Klein, D. F. 1987 Nutrients and toxins of plants in Amboseli, Kenya. *Afr. J. Ecol.* **25**, 279–293.

Barton, R. A. 1989 Foraging strategies, diet and competition in olive baboons. Ph.D. thesis, University of St Andrews.

Barton, R. A. 1992 Allometry of food intake in free-ranging primates. *Folia Primatol.* (In the press.)

Barton, R. A., Whiten, A., Byrne, R. W. & Strum, S. C. 1992*a* Habitat use and resource availability in baboons. *Anim. Behav.* (In the press.)

Barton, R. A., Whiten, A., Byrne, R. W. & English, M. 1992*b* Chemical composition of baboon plant foods: implications for the interpretation of intra- and interspecific differences in diet. *Folia Primatol.* (In the press.)

Byrne, R. W., Whiten, A. & Henzi, S. P. 1990*a* Social relationships of mountain baboons: leadership and affiliation in a non-female-bonded monkey. *Am. J. Primatol.* **20**, 313–329.

Byrne, R. W., Whiten, A. & Henzi, S. P. 1990*b* Measuring the food constraints on mountain baboons. In *Baboons: behaviour and ecology, use and care. Selected Proceedings of the XIIth Congress of the International Primatological Society* (ed. M. T. de Mello, A. Whiten & R. W. Byrne), pp. 105–122. Univ. Brasilia, Brasil.

Choo, G. 1981 Plant chemistry in relation to folivory by some colobine monkeys. Ph.D. thesis, University of Strathclyde.

Davies, A. G., Bennett, E. L. & Waterman, P. G. 1988 Food selection by two South-east Asian colobine monkeys (*Presbytis rubicunda* and *Presbytis melelophos*) in relation to plant chemistry. *Biol. J. Linn. Soc.* **34**, 33–56.

Dunbar, R. I. M. 1988 *Primate social systems.* London: Croom Helm.

Hamilton, W. J. III, Buskirk, R. E. & Buskirk, W. H. 1978 Omnivory and utilization of plant resources by chacma baboons, *Papio ursinus*. *Am. Nat.* **112**, 911–924.

Harding, R. S. O. 1981 An order of omnivores: non-human primate diets in the wild. In *Omnivorous primates: gathering and hunting in human evolution* (ed. R. S. O. Harding & G. Teleki), pp. 191–214. New York: Columbia University Press.

Johnson, R. B. 1989 The feeding strategy of adult male yellow baboons (*Papio cynocephalus*). Ph.D. thesis, University of Cambridge.

McGrew, W. C., Sharman, M. J., Baldwin, P. J. & Tutin, C. E. G. 1982 On early hominid plant food niches. *Curr. Anthrop.* **23**, 213–214.

McKey, D. B., Gartlan, J. S., Waterman, P. G. & Choo, G. M. 1981 Food selection by black colobus monkeys (*Colobus satanus*) in relation to plant chemistry. *Biol. J. Linn. Soc.* **16**, 115–146.

Milton, K. 1979 Factors affecting leaf choice by howler monkeys: a test for some hypotheses of food selection by generalist herbivores. *Am. Nat.* **14**, 362–377.

Milton, K. 1980 *The foraging strategy of howler monkeys.* New York: Columbia University Press.

Norton, G. W., Rhine, R. J., Wynn, G. W. & Wynn, R. D. 1987 Baboon diet: a five-year study of stability and variability in the plant feeding and habitat of the yellow baboons (*Papio cynocephalus*) of Mikumi National Park, Tanzania. *Folia Primatol.* **48**, 78–120.

Peters, C. R. & O'Brien, E. M. 1981 The early hominid plant-food niche: insights from an analysis of plant exploitation by *Homo*, *Pan* and *Papio* in Eastern and Southern Africa. *Curr. Anthropol.* **22**, 127–140.

Peters, C. R. & O'Brien, E. M. 1982 On early hominid plant food niches. *Curr. Anthropol.* **23**, 214–218.

Post, D. G. 1978 Feeding and ranging behaviour of the Yellow Baboon (*Papio cynocephalus*). Ph.D. thesis, Yale University.

SPSS 1983 SPSSX User's Guide. New York: McGraw Hill.

Stacey, P. B. 1986 Group size and foraging efficiency in yellow baboons. *Behav. Ecol. Sociobiol.* **18**, 175–187.

Strum, S. C. 1981 Processes and products of change: baboon predatory behaviour at Gilgil, Kenya. In *Omnivorous primates* (ed. R. S. O. Harding & G. Teleki), pp. 255–302. New York: Columbia University Press.

Waterman, P. G. 1984 Food acquisition and processing as a function of plant chemistry. In *Food acquisition and processing in primates* (ed. D. J. Chivers, B. A. Wood & A. Bilsborough), pp. 177–211. London: Plenum.

Waterman, P. G., Ross, J. A. M., Bennett, E. L. & Davies, A. G. 1988 A comparison of the floristics and leaf chemistry of the tree flora in two Malaysian rain forests and the influence of leaf chemistry on populations of colobine monkeys in the Old World. *Biol. J. Linn. Soc.* **34**, 1–32.

Whiten, A. & Barton, R. A. 1988 Demise of the checksheet: using off-the-shelf hand-held computers for remote fieldwork applications. *Tr. Ecol. Evol.* **3**, 146–148.

Whiten, A., Byrne, R. W. & Henzi, S. P. 1987 The behavioural ecology of mountain baboons. *Int. J. Primatol.* **8**, 367–388.

Whiten, A., Byrne, R. W., Waterman, P. G., Henzi, S. P. & McCulloch, F. M. 1990 Specifying the rules underlying selective foraging in wild mountain baboons, *P. ursinus*. In *Baboons: behaviour and ecology, use and care. Selected Proceedings of the XIIth Congress of the International Primatological Society* (ed. M. T. de Mello, A. Whiten & R. W. Byrne), pp. 5–22. Univ. Brasilia, Brasil.

Discussion

K. Hawkes (*Department of Anthropology, University of Utah, Salt Lake City, U.S.A.*). What about the costs of these foods? Optimal foraging models predict choices based on the nutrient payoff (often measured as calories) per unit of time spent foraging. The 'optimal diet' or 'prey choice' model predicts the inclusion of resources if their profitability (the rate earned for pursuing and handling them after they have been found) is at least as high as the expected rate for passing them by and continuing to search for something better. If the resources exploited by human foragers are evaluated only from the point of view of nutrients per kilogram, many with very low values are exploited whereas many with much higher values are not. When handling costs are considered, and resources are evaluated for nutrients gained per unit time, the economies can be quite clear. Could it be that Dr Whiten's anomalous points are resources with very different handling costs?

A. Whiten. It seems unlikely that the anomalous point in figure 2 can be explained in this way, as this corresponds to an unripe phase of a fruit eaten when ripe and the two thus have similar harvesting and handling properties. However, Dr Hawkes' general point about taking account of costs is an important one. Considering cost–benefit equations is one reason for our multivariate approach, and constituents with negative coefficients can be regarded as one type of cost. The additional costs Dr Hawkes emphasizes can be divided into time-to-find and time-to-process. With respect to the latter, that is something we cannot take precise account of in contrasts of food species with species that are not eaten and therefore not processed at all. However, we took care to make paired contrasts (e.g. eaten deep corm with non-eaten deep corm) that minimize likely handling time differences. In contrasts between parts of a plant eaten with those discarded, it is often the part eaten that takes longest to prepare, so the preference cannot be explained by time constraints. However, time-to-process is a variable which it will make sense to include in our further research on dietary preferences within the set of foods consumed.

Time-to-find is a more elusive quantity to measure because nobody has discovered a way to define baboons' search strategy. If we assumed random search, we can at least consider abundance of alternative items. Again, this is automatically taken account of in contrasts between parts discarded or consumed, and in contrasts with species not eaten we have selected items that are far from rare. In examining preferences among the set of food items, abundance is explicitly taken account of in our selection ratios.

E. M. Widdowson (*9 Boot Lane, Barrington, Cambridge, U.K.*). Can Dr Whiten explain how a baboon discriminates for protein and against fibre? Neither has any taste or smell. Does the baboon try the non-food and dislike it, or is the discrimination learned and passed on from generation to generation?

S. A. Altmann (*Department of Ecology and Evolution, University of Chicago, Illinois, U.S.A.*). The characteristics that a discriminant function uses to discriminate between foods and non-foods may not be the ones that the animals use. Much work has been done on the basis for diet selection in mammals. Initially, the research on specific hungers, initiated by Curt Richter (1943), suggested that these inborn physiological mechanisms might provide the basis for selecting nutritionally adequate diet. However, subsequent research has shown that the number of these specific hungers is very limited. They have been documented for sodium, water, energy, oxygen and possibly proteins, but not for any other nutrients (Rozin 1976). Consequently, specific hungers cannot be relied on for obtaining a balanced diet, unless, perchance in a particular habitat, an adequate intake of other nutrients is entailed by their covariance with these basic ones. Considerable research has been done on the behavioural mechanisms by which mammals obtain an adequate diet in the absence of a specific hunger for each nutrient (Barker *et al.* 1977), but many aspects of this complex problem remain to be solved. In this area, we can have a fruitful interchange of ideas between those doing experimental laboratory studies and those studying the natural diets of wild animals.

References

Barker, L. M., Best, M. R. & Domjan, M. (eds) 1977 *Learning mechanisms in food selection.* Baylor University Press.

Richter, C. 1943 Total self regulatory functions in animals and human beings. Harvey Lect. **38**, 63–103.

Rozin, P. 1976 The selection of foods by rats, humans, and other animals. *Adv. Study Behav.* **6**, 21–76.

K. Milton (*Department of Anthropology, University of California, Berkeley, U.S.A.*). I have an additional comment on Dr Widdowson's question to Dr Whiten. From my work with feeding behaviour of *Hapalemur griseus*, a bamboo-specialist and *Alouatta palliata*, a folivore, I noted that individuals of both species sniffed foliage and then instantly either accepted or rejected it. This suggests that there may be olfactory cues that both prosimians and anthropoids can utilize to differentiate edible from non-edible foliage.

A. Whiten. The answer to Dr Widdowson's specific question about baboons is that we do not know. However, expanding on Professor Altmann's comment, it may be helpful at this stage of research to list the most obvious candidates for proximate mechanisms of selection or avoidance of specific constituents like fibre and protein, and summarize some relevant data which are available for other species. I suggest four alternative mechanisms: (i) direct perception through, for example, taste, smell and texture; (ii) a constituent might instead have after effects (e.g. satiety, relief of symptoms of a particular nutrient deficiency), which the animal learns to associate with immediate perceptual cues of the food concerned (and of course such learning could operate with respect to (i) also, as in learning to recognize by sight a food which tastes good); (iii) animals may evolve an innate liking for the sensations of particular foods or food classes which in their habitat offer a beneficial mixture of constituents; or (iv) young animals may copy the food habits they observe in others (although this last mechanism still begs the question of how such a cultural preference would be established in the first place).

Considering each of these possibilities with respect to protein harvesting, I know of no direct evidence bearing on (iii). By contrast there is a significant literature on possibilities (i) and (ii). This is based mostly on extensive experiments with rats (e.g. Baker *et al.* 1987), but K. Milton has drawn my attention to one experiment by Peregoy *et al.* (1972) which used monkeys. After rearing from 120–210 days old on a low (3.5%) protein diet, rhesus monkeys were allowed to feed from bins of three differently coloured foods with protein contents of 2%, 3% and 25%. Unlike well-nourished control animals, the protein depleted monkeys expressed a preference for the high protein diet within nine minutes of starting to feed. This preference became stronger over succeeding days. When the colours of the 2% and 25% diets were then reversed, the low-protein monkeys at first continued with their (now low-protein) colour preference, but over a period of days learned to switch colour preference, again selecting the 25% protein diet.

Both the rat and monkey experiments have generally been assumed to be evidence for mechanism (ii), although the fact that the monkeys studied by Peregoy *et al.* expressed a preference within nine minutes offers some support for mechanism (i). Mechanism (i) has now been advocated by Deutsch *et al.* (1989), who showed protein-deprived rats' immediate preferences for a number of specific proteins such as gluten and ovalbumin, which could not have been learned before the tests. There is evidence that these unlearned preferences are based on olfactory cues (Heinrichs *et al.* 1990). Wallin (1988) has further examined genetic components in such choice behaviour, artificially selecting for preferences for high protein content or for high energy content through 44 generations of fruitfly larvae. Both lines developed higher preferences in the selected direction.

Although there is evidence that primates readily learn aspects of foraging by watching others (mechanism (iv); Whiten 1989; Whiten & Ham 1992), the work of Peregoy *et al.* suggests that if a monkey was (strangely) led through observation of others initially to select a protein-poor diet, it would be capable of quickly and adaptively adjusting its preferences through individual learning.

I know of no equivalent work on fibre selection, but possibilities (i) and (ii) seem more obviously plausible in this case.

References

Baker, B. J., Booth, D. A., Duggan, J. P. & Gibson, E. L. 1987 Protein appetite demonstrated: learned specificity of protein-cue preference to protein need in adult rats. *Nutr. Res.* **7**, 481–487.

Deutsch, J. A., Moore, B. O. & Heinrichs, S. C. 1989 Unlearned specific appetite for protein. *Physiol. Behav.* **46**, 619–624.

Heinrichs, S. C., Deutsch, J. A. & Moore, B. O. 1990 Olfactory self-selection of protein containing foods. *Physiol. Behav.* **47**, 409–413.

Peregoy, P. L., Zimmerman, R. & Strobel, D. A. 1972 Protein preference in protein-malnourished monkeys. *Percept. Motor Skills* **35**, 495–503.

Wallin, A. 1988 The genetics of foraging behaviour: artificial selection for food choice in larvae of the fruitfly, *Drosophila melanogaster*. *Anim. Behav.* **36**, 106–114.

Whiten, A. 1989 Transmission mechanisms in primate cultural evolution. *Trends Ecol. Evol.* **4**, 61–62.

Whiten, A. & Ham, R. 1992 On the nature and evolution of imitation in the animal kingdom: reappraisal of a century of research. In *Adv. Study Behav. 21* (ed. P. J. B. Slater, J. S. Rosenblatt, C. Beer & M. Milinski). New York: Academic Press. (In the press.)

R. I. M. Dunbar (*Department of Anthropology, University College London, U.K.*). Most of the discussion of dietary nutrients has focused on proteins and secondary compounds, and the role of energy in animals' nutritional budgets has been ignored. Yet, analyses of the time budgets of baboons shows that a significant proportion of the time that animals in a given population have to spend feeding is a function of the habitat's ambient temperature, suggesting that thermoregulatory costs may be important. For a tropical mammal, this in itself is surprising, as it seems that baboons face significant thermoregulatory costs even at altitudes as low as 1000 m above sea level. But it does seem to me to underline the importance of energy as a component of an animal's diet.

A. Whiten. I agree with Dr Dunbar that available energy should be an important 'constituent' to consider and in principle this could be included in the type of multivariate analysis we have advocated. It may turn out to discriminate foods and non-foods even better than the specific nutrients we have considered. However, we have been reluctant to assess overall energy by using calorimetry, because we do not know how much of this energy can be extracted (wood, after all, is a non-food with high energy content). We have preferred to examine specific nutrients, some of which are obvious candidates for selection as energy sources. Of these, fat was shown to be positively valued, particularly in the class of fleshy stem bases which constituted the staple food in the winter period. Starch, by contrast, which was common in some storage organs, did not appear to figure in selection rules: nor did total carbohydrate. However, the latter was estimated crudely, by subtraction. Once we can determine digestibile carbohydrates, it will be possible to compute available energy from this and the protein and fat components, and enter this into the multivariate analyses.

Hominoid dietary evolution

PETER ANDREWS[1] AND LAWRENCE MARTIN[2]

[1] *Natural History Museum, Cromwell Road, London SW7 5BD, U.K.*
[2] *Departments of Anthropology and Anatomical Sciences, State University of New York, Stony Brook, 11794, U.S.A.*

SUMMARY

During the later Palaeocene and early Miocene, catarrhine primates and the evolving hominoids had adaptations for frugivorous diets, with the emphasis on soft foods. Early in the middle Miocene the hominoids underwent a major shift, both in morphology and in habitat, with the morphology characterized by thickened enamel on the molars, enlarged incisors and massive jaws. The diet indicated by this morphology is interpreted as still mainly frugivorous but with changed emphasis, possibly towards harder objects. The thick-enamelled hominoids are found associated with more open forest habitats, and the distribution of food resources in equivalent habitats today is discontinuous both in time and in space, leading to evolutionary pressures particularly affecting locomotion, brain size and social behaviour. The earliest known hominid fossils differed little in dental and mandibular morphology from the middle Miocene apes, and the implied dietary similarity, together with ape-like patterns of dental development and retained arboreal adaptations of the postcrania, suggests little change in the foraging strategies of the earliest hominids compared with their ape ancestors and further suggests similarity in evolutionary grade. This similarity may have extended to other aspects of behaviour, for example to patterns of tool making and use, which may have been similar in the common ancestor of apes and humans to the pattern shared by the earliest australopithecines and chimpanzees.

1. INTRODUCTION

Hominoid primates are known from the Miocene to the Present. Relationships among extant hominoids are resolved with some degree of certainty, but the timing and location of most evolutionary events leading to the living apes and humans are not known precisely. We are taking as a starting point the phylogeny that we have formulated in a series of earlier papers (Andrews 1978, 1985; Martin 1983, 1985; Martin & Andrews 1984; Andrews & Martin 1987*a*, *b*). The phylogeny shown in simplified form in figure 1 is based on two sources of evidence: the branching pattern of the extant hominoids, giving their evolutionary relationships, was first established by reference to recent taxa only; the timing of the evolutionary events, and the fine detail of morphological change, was then interpreted by fitting fossil hominoids into the pattern already established. We propose to follow the same model in interpreting the nature of dietary change in the course of hominoid evolution.

By necessity we will be using adaptationist models for inference of diet, as morphology provides the only positive source of dietary information for fossil animals by analogy with the morphology–diet correlations observed in living primates. In this regard, however, it is important to distinguish what might be called primary adaptation from heritage characters, the former being a direct response to environmental conditions, improving the fitness of individuals relative to others lacking the adaptation, whereas the latter are retained unchanged from an ancestor, and may not be directly linked with changing diet. This distinction may be important in interpreting morphological change in fossil apes, although it has been found that living ape morphology is more closely linked to dietary function than to phylogenetic history (Hartman 1988).

Morphology provides our principal source of evidence from fossils, although palaeoecology and dental microwear are also important. All of these, however, reflect only the principal (and probably most recent) diet of the species in question, whether recent or fossil, and they cannot at present take account of elements that constitute minor parts of all hominoid diets (Ghiglieri 1987). By these criteria, it may be concluded that hominoid primates are a mainly frugivorous group (see Tutin *et al.*, this symposium), and this feeding category can be further subdivided into 'hard' and 'soft' groups: hard food includes seeds, nuts and unripe fruit, and may also include non-fruit items such as hard-carapaced insects; soft food includes ripe fruit and soft-bodied insects (Lucas & Luke 1984). Both categories can be distinguished from other food categories such as tough (leaves and animal flesh), or large and hard (bones).

2. MORPHOLOGICAL CORRELATES OF DIET

In most groups of mammals, dental morphology is correlated with diet at a general level. In this paper we provide evidence for enamel thickness for all stages of hominoid evolution (Martin 1983, 1985), and this will

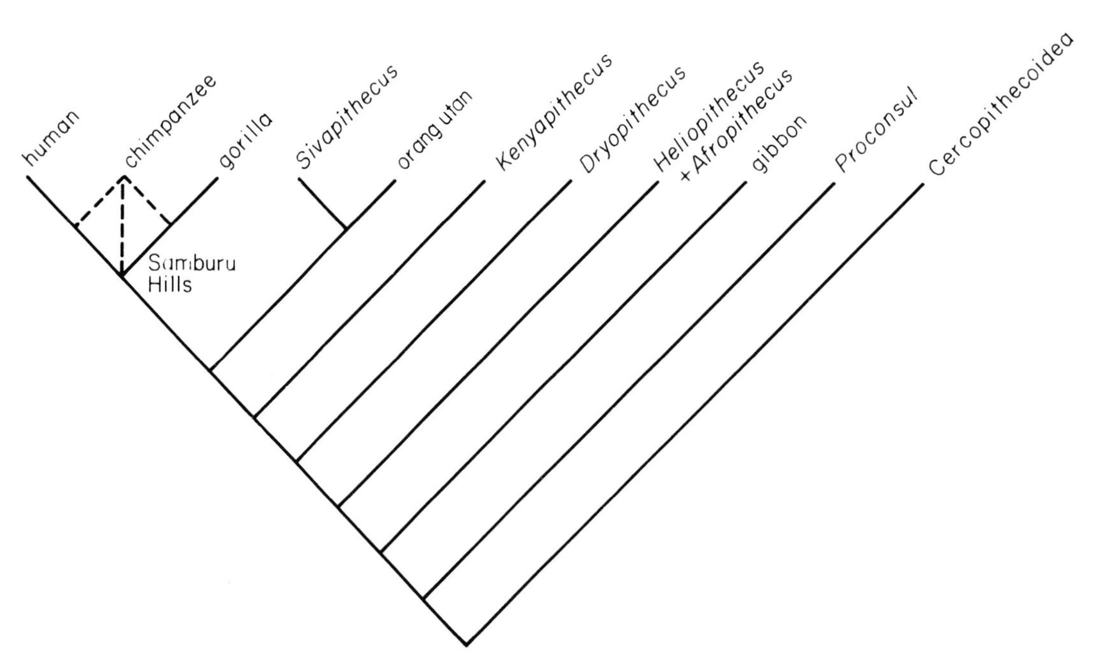

Figure 1. Cladogram showing relationships of fossil and recent hominoids.

be supplemented by other lines of evidence where available. Early work on enamel thickness based on worn teeth and natural breaks in teeth provided some insights into variations in enamel thickness (Kay 1981; Beynon & Wood 1986), but we base our present work on the physical sectioning of teeth and measuring the cross-sectional area of the enamel (Martin 1983). This area is divided by the length of the enamel–dentine junction to produce an average thickness, and further divided by the square root of the dentine area exposed in the same section to derive a size-standardized thickness. It might be argued that absolute thickness of enamel is the appropriate measure in relation to feeding studies, and we would agree if rates of tooth wear were being considered, but we maintain that relative enamel thickness is more useful for comparisons between animals differing in size, masticatory forces and longevity (Martin 1985).

Other lines of evidence, unfortunately still far from complete, are as follows.

1. Body size can be related to diet in a general way. There is progression upwards in size from insectivory in small animals to herbivory in large animals, with frugivory in between (Walker 1981; Andrews 1983). The largest living primate that has an insectivorous diet has a mass of no more than 250 g, and, in the animal kingdom as a whole, the few insectivorous species larger than this have developed highly specialized collecting strategies for diets of ants or termites. Conversely, the smallest living folivorous primate has a mass of 700 g, and, for energetic reasons relating to digestion times where cellulose is the main food source, most folivores are considerably larger.

2. Robusticity of the mandible is related to both diet and body size. The relation is complex, however, and regressions have been calculated separately both for the three main groups of higher primate and for males and females (Alcock 1984). Residuals from these lines were then calculated, but none were found to be significant at the 95% level.

3. Incisor size relative to body size (or its correlate, M1 size) is indicative of the type of food processing. Frugivores have relatively small molars and large incisors, whereas in folivores the proportions are reversed. The ratio of incisors to molars differs markedly, therefore, for the two dietary types.

4. Crown height and molar occlusal area are potentially important attributes of molar teeth, but little comparative information exists for primates. Australopithecines are said to be megadont, and large species have relatively larger teeth for their size than do smaller species, but this appears likely to be a unique character of this group and its significance is disputed (Kay 1985).

5. The lengths of the shearing crests on the occlusal surfaces of molars have been combined in a shearing quotient (Kay 1977; Maier 1984). Long shearing blades may be combined with high crowns and thin enamel, and this combination is characteristic of diets with high fibre content, such as folivory. In this morphology, advancing wear maintains or increases the shearing blades, whereas for thick-enamelled teeth with little cusp relief, advancing wear flattens the occlusal relief.

6. Patterns of microwear can be observed through differences in numbers and sizes of wear striations, as well as relative proportions of pits as opposed to striations (Walker *et al.* 1978). These patterns can be related to a number of dietary differences in living animals, such as browsing against grazing, forest against open-country herbivory, feeding on vertebrates against invertebrates (Teaford 1988). The patterns show a high degree of regularity, but non-dietary factors have also been found to influence them (Gordon 1982).

7. Some aspects of positional behaviour are closely associated with feeding. Below-branch arm hanging is part of a greater range of positional behaviour that has been observed in apes and is associated with small terminal branch feeding (Aiello 1981). Hand function

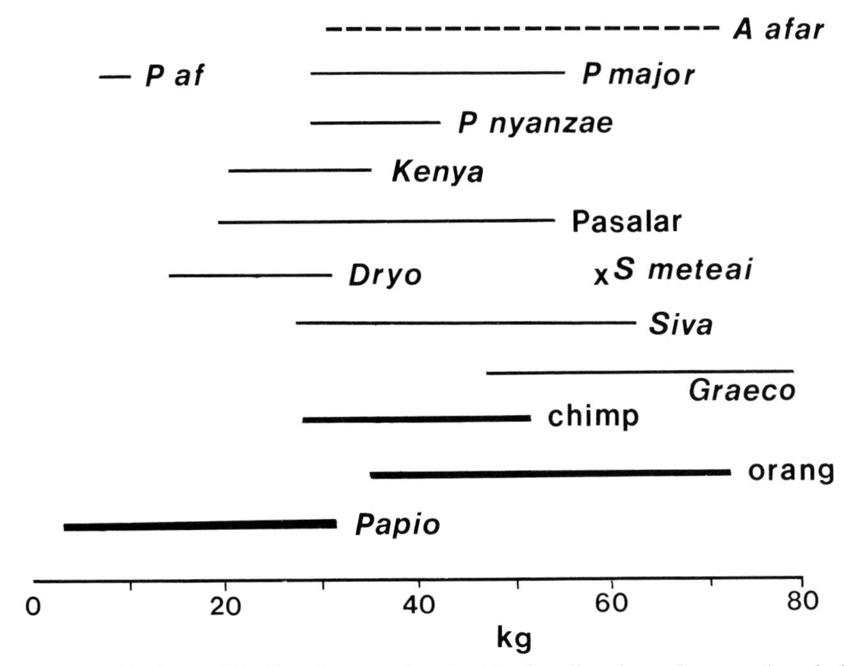

Figure 2. Body mass distributions of fossil and recent hominoids. Reading from the top, *Australopithecus afarensis*, *Proconsul africanus*, *P. major*, *P. nyanzae*, *Kenyapithecus africanus*, the sample of '*Sivapitheus*' from Paşalar, *Dryopithecus fontani*, *Sivapithecus meteai*, *Sivapithecus sivalensis*, *Graecopithecus freybergi* from Greece (= '*Ouranopithecus macedoniensis*'), chimpanzee *Pan troglodytes*, orang utan *Pongo pygmaeus*, *Papio cynocephalus*. Body mass of recent taxa and *Proconsul* from Ruff *et al.* (1989), and the others calculated from the regression of M_3 on body mass; log 10 body mass = 2.86 log M2 length + 1.37.

and degree of manipulative ability are also important factors, and in apes this is closely related to thumb length (Marzke 1983). Manipulative ability is also a significant factor in the use of tools for food processing, and this needs to be considered not just for later hominids but also for earlier stages of hominoid evolution.

3. EVOLUTIONARY TRENDS IN PRIMATE DIETS

Dietary interpretations will be reviewed chronologically, based on the results given in tables 1–3 and figures 1–4.

(*a*) *Eocene and Oligocene*

During much of the early stages of primate evolution, primates were small insect-eating animals. During the Eocene, some primates increased in body size, from an average of well below 1 kg in the early to early middle Eocene to an average slightly above 1 kg in the late Eocene and Oligocene, and at the same time there was a shift from primarily insectivorous diets in the early Eocene at about 50 Ma BP, to primarily frugivorous in the middle Eocene at about 45 Ma BP, to primarily herbivorous by the late Eocene 35–40 Ma BP (Hooker 1991). The Eocene to Oligocene primates of the Fayum showed a variety of diets (Kay & Simons 1980), and molar morphology and lack of shearing blades on the teeth of *Propliopithecus* and the other hominoid-like species known from the Fayum suggest soft-tissue diets, probably of fruit (Kay & Simons 1980). These taxa are close to the ancestry of Old World monkeys and apes, that is, to the Catarrhini, and they show great morphological and functional similarity to the apes. It is likely, therefore, that the ancestral pattern for catarrhine diets was some form of soft-fruit frugivory.

(*b*) *Early Miocene*

Two groups will be discussed from early Miocene deposits, the *Proconsul* group dating from 18 to 20 Ma BP, and the *Afropithecus* and *Heliopithecus* group from 17 to 18 Ma BP. *Proconsul* is recognized as a stem hominoid (figure 1), part of the superfamily Hominoidea but not related to any one part of the monophyletic clade (Andrews 1985; Martin 1986). Most of the genera in this group of primates are associated with tropical rain forest habitats (Andrews *et al.* 1979). They were arboreal quadrupeds living in the upper canopies much like New World monkeys today (Napier & Davis 1959; Walker & Teaford 1989). The other group combines a different set of ancestral and uniquely derived characters that places it at the stem of the great ape and human clade – Hominidae of some authors (Andrews 1985) – but its ecological affinities are not clear. *Dryopithecus* appears to be a later European derivative of this group.

(i) *Proconsul* faunas. During the early part of the early Miocene in East Africa, there was an abundance of hominoid and hominoid-like primate species. At several sites there were five to six hominoid species commonly present (Andrews 1978; Harrison 1988) together with three to six non-hominoids, with a size distribution similar to African primate faunas today (Fleagle 1978).

Table 1. *Relative enamel thickness in fossil and recent hominoids*

taxon	$(c/e)/\sqrt{b}$	s.d.	n	category
Proconsul africanus[c]	8.54	—	1	thin
Gorilla gorilla[a]	10.04	1.74	17	thin
Pan troglodytes[a]	10.10	2.09	14	thin
Hylobates lar[a]	11.02	—	1	thin
Dryopithecus fontani[c]	12.74	—	1	int. thin
Proconsul major[c]	12.84	—	1	int. thin
Oreopithecus bambolii[c]	15.46	—	1	int. thick
Pongo pygmaeus[a]	15.93	2.51	17	int. thick
Heliopithecus leakeyi[d]	17.35	—	1	int. thick
Paşalar hominoids[a]	19.71	2.49	6	thick
Sivapithecus[a]	19.73	(3.33)	3	thick
Australopithecus africanus[b]	22.17	—	2	thick
Homo[a] *sapiens*	22.35	6.23	13	thick
Graecopithecus freybergi[c]	28.34	—	1	thick–hyper thick
Paranthropus crassidens[b]	29.61	—	1	hyper thick
Paranthropus robustus[b]	31.32	—	1	hyper thick
Paranthropus boisei[b]	34.91	—	2	hyper thick

[a] Martin (1983).
[b] Grine & Martin (1988).
[c] New data.
[d] Andrews & Martin (1987*b*).

Table 2. *Comparison of mesiodistal lengths of upper central incisors with first upper molars for recent and fossil hominoids* (I^1/M^1); *n is the sample size.*

taxon	mean	range	n
orang utan	1.13	0.95–1.22	16
chimpanzee	1.09	0.95–1.24	8
gorilla	0.90	0.84–0.96	12
Proconsul	0.93	0.90–0.98	3
Afropithecus[a]	0.98	—	1
Sivapithecus[b]	0.98	0.98–0.99	2
Graecopithecus[c]	0.84	0.82–0.87	2
Dryopithecus	0.81	—	1
Australopithecus africanus[d]	0.81	0.76–0.89	3
A. afarensis[d]	0.89	0.86–0.93	2
Paranthropus boisei	0.60	—	1

[a] Leakey & Leakey (1986).
[b] Includes '*Ankarapithecus*' (Martin & Andrews 1984).
[c] de Bonis *et al.* (1990).
[d] White (1977).

Enamel thickness data are reported here for the first time for *Proconsul africanus* and *P. major*. *P. africans* has thin enamel (table 1), which is the inferred ancestral condition for the hominoid clade (Martin 1985; Andrews & Martin 1987*a*), while *P. major* has slightly thicker enamel. Incisor size (table 2) in most Miocene species is slightly larger than the condition interpreted as primitive for the Hominoidea (Andrews 1985). The I1 is also more spatulate, as in living hominoids, and both features are indicative of fruit processing. Shearing quotients are mostly low, suggesting non-fibrous diets (Kay 1977). They group around values for gibbons, chimpanzees and frugivorous New World monkeys, and some form of soft-fruit frugivory would therefore appear to be indicated for *Proconsul* species and *Limnopithecus*. *Rangwapithecus gordoni* has both more cuspidate teeth and a higher shearing quotient than other early Miocene hominoids, and on these grounds it may be presumed to have been more folivorous.

The morphology of the hand in *Proconsul africanus* contrasts with the living apes in retaining a primitively longer thumb. It is approximately 50% of the length of the third digit, the same as in modern humans. The angle of the trapezium was at least as great as in chimpanzees (Napier & Davis 1959), so that, in combination with the longer thumb, the fossil ape may have had manipulative ability exceeding that of living apes. The combination of many primitive characters of the hand must make any functional inferences very tentative (Beard *et al.* 1986), but manipulative ability in the hand of *Proconsul* may have been closer to australopithecines than to extant apes.

(ii) *Afropithecus* faunas. Towards the end of the early Miocene there occurred a recently described group of hominoid primates from East Africa and the Arabian peninsula. *Afropithecus turkanensis* was a large species, with massive jaws and robustly built skull. It had enlarged incisors (table 2) and molars with low rounded cusps (Leakey & Leakey 1986). *Heliopithecus leakeyi* from Arabia is similar to it in both age and morphology, but is considerably smaller. The enamel is of intermediate thickness (Andrews & Martin 1987*b*), thicker than in the living African apes and *Proconsul*, and at the top end of the range for orang utans (table 1). The significance of thicker enamel will be discussed in the next section, but the combination of characters seen in *Afropithecus* and *Heliopithecus* leaves little doubt that fruits were a major part of their diet.

(*c*) **Middle Miocene**

Thick-enamelled hominoids, such as *Kenyapithecus* from East Africa and '*Sivapithecus*' from Pasalar and

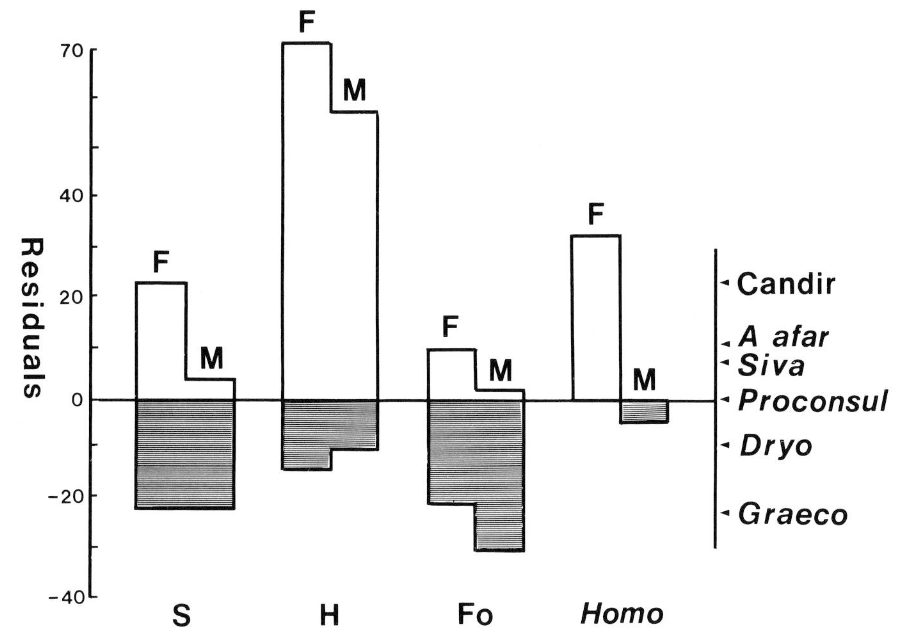

Figure 3. Mandibular robusticity based on residuals from the regression lines of horizontal section modules of the mandible on an estimate of body size – the area of the molar tooth row (Alcock 1984). Separate regressions were calculated for New World monkeys, Old World monkeys and hominoids. Results are for '*Sivapithecus*' from Candir, *Australopithecus afarensis*, *Sivapithecus*, *Proconsul*, *Dryopithecus*, *Graecopithecus*; S, anthropoids with soft fruit diet; H, anthropoids with hard fruit diet; FO, anthropoids with folivorous diet; M, male; F, female.

Candir in Turkey, both dating to about 15 Ma BP, first appeared in the middle Miocene. They are seen as a continuation of the early Miocene trend towards thickening enamel initiated in the progression from *Proconsul* to *Afropithecus* and *Heliopithecus*. They were associated with more open habitats than were the early Miocene hominoids: tropical to subtropical deciduous forest is indicated for both (Andrews *et al.* 1979; Andrews 1990), with a strongly seasonal climate and hence seasonal supplies of fruit. Derived from this group are the sivapithecines, now seen as related to the orang utan clade (Andrews & Cronin 1982; Ward & Pilbeam 1983) and including, for our purposes here, the Greek genus *Graecopithecus* (Martin & Andrews 1984).

(i) *Kenyapithecus* faunas. The fossil hominoids from the middle Miocene of Kenya and Turkey are characterized by thick enamel on their molar teeth. Thick enamel can only be inferred for *Kenyapithecus* based on radiographs and broken teeth, but a sample of the Turkish teeth has been sectioned (table 1, figure 4). The set of characters shared by *Kenyapithecus* and the Turkish '*Sivapithecus*' also includes molars with flattened occlusal surfaces, especially after some wear, robust but still single-cusped P_3s, and robust maxillary and mandibular bodies (figure 3). The maxillary body is deep with wide separation between the alveolar border and the floor of the maxillary sinus, and the mandible has a massively elongated symphysis, much longer than deep (Alpagut *et al.* 1990). No associations are known of incisors with molars, but based on means of isolated teeth their size relative to the molars appears little different from that seen in *Proconsul*, but there is an indication that the upper central incisors are much larger than the laterals, a characteristic of the orang utan clade.

The postcranial remains of *Kenyapithecus* suggest generalized arboreal quadrupedalism in this genus (Rose 1983) of a type similar to that of *Proconsul* but with no exact parallel today. No hand bones are known.

(ii) Orang utan lineage. Thick enamel is present also in the hominoids that are phylogenetically linked with the orang utan, such as *Sivapithecus* and its included genus *Ramapithecus*. New data presented here suggest the presence of even thicker enamel in *Graecopithecus* (table 1). The presence of thick enamel at this stage would appear to be a retained character from a thick-enamelled ancestor of the middle Miocene of Kenya or Turkey, and thick enamel may have either retained its adaptive function in *Sivapithecus*, or be a heritage character whose function has changed. There is, however, some additional evidence for *Sivapithecus* from the microwear on the molars. The relative proportions of pits and scratches, as well as lengths and breadths of defects, on *Sivapithecus* molars match the soft-fruit pattern and is similar to that observed today in chimpanzees (Teaford & Walker 1984). It is significantly different (Teaford 1988) from the pattern observed in animals that eat hard fruit (which is the function assigned by Kay (1981) to thick enamel). Teaford & Walker (1984) conclude from this that the function of thick enamel is uncertain, but it is also possible that its function when it first evolved was, as Kay suggested, for grinding hard objects, but later hominoids changed diet to softer fruits as in chimpanzees without at first losing thick enamel. Study of the microwear of *Heliopithecus* or *Kenyapithecus* would help to resolve this issue.

The upper central incisors of *Sivapithecus* are robust teeth with massive lingual cingula. The same tooth in *Graecopithecus* is extremely low crowned (de Bonis *et al.*

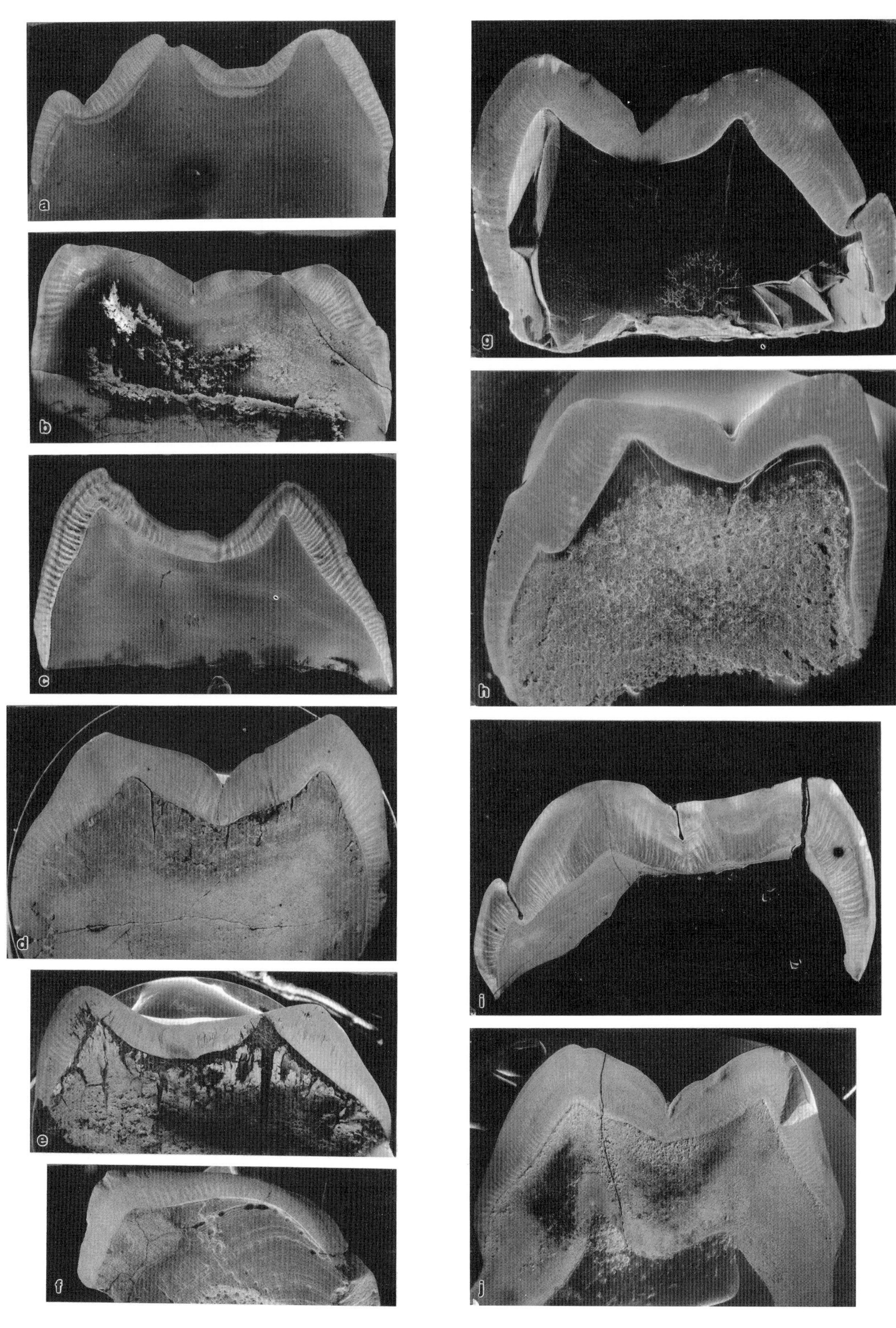

Figure 4. For description see opposite.

1990), and it is smaller relative to first molar length (table 1). In both cases the central incisor is very much larger than the lateral incisor, a specialization seen in living orang utans but not in any other group, living or fossil. Although it would appear to be an important adaptation for fruit processing, the differing proportions from all extant hominoids makes interpretation difficult. Postcranially, *Sivapithecus* was functionally similar to extant cebids, generalized quadrupedalism with a varied positional repertoire and none of the specializations of living apes or monkeys (Rose 1989).

(*d*) *Late Miocene and Pliocene*

Compared with the early stages of hominoid evolution, and in particular with the origin of the orang utan lineage, there is practically no fossil evidence for these periods. This is all the more regrettable in that it covers the time of human and African ape divergence (figure 1). Eleven fragmentary fossils are known prior to 4 Ma BP (Hill & Ward 1988), some of them probably pre-dating the splitting of the human lineage. Dental specimens retain thick molar enamel, judging from the external appearance of the teeth, but no measurements have been made. Thick enamel even appears to be characteristic of the Samburu Hills maxilla (Hill & Ward 1988), which we consider to be phylogenetically linked with the gorilla, and this is partial confirmation of Martin's (1983) prediction (Andrews & Martin 1987*a*) of thick-enamelled ancestry for the African apes. The mandibles from Tabarin and Lothagam, however, are probably early members of the hominid clade. They have low robust mandibular bodies (figure 3), and in size and morphology they closely resemble *Australopithecus afarensis* from Laetoli (White 1977).

A. afarensis has a pattern of microwear on the anterior teeth that has a combination of gorilla-like and unique features. Even when the P_3 is single-cusped and the canines high crowned and projecting, they are not used in the same way as in great apes to slice and shear food, but rather they function to puncture–crush food. Moreover, the presence of both polishing, and pitting and micro-flaking, on the incisors suggests food procurement from different sources (Ryan & Johanson 1989). The proportions of the incisors of *A. afarensis* appear primitive (table 2), and the molars appear to retain primitively thick enamel. There is no evidence from the molar or mandibular morphology of any major dietary change from the pattern observed in thick-enamelled apes during the preceding 10 Ma. Most interpretations of Laetoli and Hadar palaeoenvironments, where *A. afarensis* has been discovered, suggest either open savanna habitats (Leakey & Harris 1987) or more wooded habitats with seasonally deciduous woodland or forest (Andrews 1989). There are also indications that the hominids retained a considerable arboreal capacity (Stern & Susman 1983). Later species of australopithecine have extensive modifications of their teeth, particularly the robust australopithecines (Grine 1986; Kay & Grine 1988; Grine & Martin 1988), but these will not be discussed here.

4. HOMINOID DIETARY TRENDS

Several trends have become apparent from the foregoing descriptions of hominoid tooth morphology. The most notable of these was the increase in enamel thickness during the early to middle Miocene, coupled with the development of flattened molar occlusal morphology, and these were often accompanied by premolar enlargement and increase in mandibular robusticity, as seen in the sequence: *Proconsul–Heliopithecus–Kenyapithecus*. The dietary significance of these changes is far from certain, as the functional significance of thick enamel is still being debated. It has been claimed to be associated with terrestrial foraging (Jolly 1970) as an adaptation for hard abrasive foods found on the ground, but Kay (1981) rejected the terrestrial aspect of this idea, claiming that hard and brittle foods alone may have provided the selective impetus for thick enamel, regardless of where they occurred. Kay (1984) also found that enamel thickness varies inversely with shearing lengths, and as wear proceeds the occlusal surfaces of thick-enamelled teeth become smoothly rounded. Small but hard objects make up a large part of the diet of the orang utan, which also has moderately thick enamel, whereas the folivorous gorilla has the

Figure 4. Coronal sections through the mesial cusps of fossil hominoid molars. Block faces revealed by a wire saw cut through the mesial cusps in a coronal plane. Faces were oriented perpendicular to the incident electron beam in an AMR 1810 Scanning Electron Microscope and imaged using secondary electrons. The images in this plate have been printed to a magnification such that all of the teeth have approximately equal cervical diameters, i.e. the pictures are scaled to a constant tooth size rather than being printed at a uniform magnification. Fieldwidths are given separately for each image and are the width of the entire image field. (*a*) *Proconsul africanus* right M^1 (M 14081) mesial posterior face, fieldwidth = 8.62 mm. (*b*) *Proconsul major* left M_3 (M 32237) mesial anterior face, fieldwidth = 13.65 mm. (*c*) *Dryopithecus fontani* right M^1 (IPS 68) mesial anterior face, fieldwidth = 10.6 mm. (*d*) *Sivapithecus sivalensis* right M^1 (M 13366) mesial anterior face, fieldwidth = 12.94 mm. (*e*) *Sivapithecus punjabicus* left M_3 (M 13367) mesial anterior face, fieldwidth = 11.92 mm. (*f*) *Heliopithecus leakeyi* right M^3 (M 35146). Fracture face of the lateral paracone, which has been polished to smooth the rough, fractured surface. The fracture passes through the tip of the paracone, but runs somewhat obliquely relative to both coronal and sagittal planes so that the plane runs somewhat distally as it moves towards the cervix. Fieldwidth = 8.45 mm. (*g*) Paşalar hominoid right M_1 (BP 13) mesial anterior face, fieldwidth = 9.08 mm. (*h*) Paşalar hominoid left M_1 (BP 14) mesial anterior face, fieldwidth = 7.64 mm. (*i*) *Graecopithecus freybergi* right M_1 (RPL 641) mesial posterior face, fieldwidth = 16.59 mm. (*j*) *Australopithecus africanus* left M^2 (Stw 284) mesial anterior face, fieldwidth = 16.82 mm.

thinnest enamel and the longest shearing crests seen in extant apes (Kay 1977; Martin 1985). Similarly, with monkeys, the species with diets of hard fruit have the thickest enamel, whereas folivorous species have the thinnest enamel (Kay 1981, 1985).

The dietary category of ancestral catarrhines and earliest hominoids is widely interpreted as soft-fruit frugivory, and the trend towards increasing enamel thickness probably suggests a shift to hard-fruit frugivory rather than any substantive change in diet. The evidence of microwear on later Miocene thick-enamelled hominoids is most similar to that seen today on soft-fruit eaters (Teaford & Walker 1984), which suggests no change from their thin-enamelled ancestors, but no work has yet been done on such fossils as *Heliopithecus* and *Kenyapithecus* where the trend of increasing enamel thickness is first observed. Diets of the later sivapithecines may have changed, as indicated by the microwear, without apparent change of the enamel, and it is on the teeth of the fossil hominoids where thick enamel is first observed that the significance of thick enamel is most likely to be understood.

Another interpretation of thick enamel is that it was an adaptation for a more varied diet. This would have been the product of increasingly seasonal climates and habitats, the seasonality being in part the result of topographic change within the tropics, producing more open and seasonally varied habitats at *Kenyapithecus* localities such as Fort Ternan and Maboko Islands (Andrews *et al.* 1979), and in part latitudinal, as hominoids spread northwards beyond the tropics (Andrews 1990). Chimpanzees living today in seasonal habitats have been observed to have more varied diets than forest chimpanzees (Collins & McGrew 1988), and during the dry season they eat dry fruits, hard seeds and hard-shelled fruits. Even so, they are restricted to areas of mosaic habitat, depending on locally wetter areas, such as gallery forest, during the dry season. They occur at lower population densities, and they move about their enlarged home ranges to a greater extent; here their increased body size is an advantage, giving them greater mobility and the capacity to feed on larger food items than possibly competing cercopithecoid monkeys (Wrangham 1980; Andrews 1981). In this scenario, thick enamel is coupled with increased body size in seasonal mosaic habitats.

A further possibility is that thick enamel did not evolve in response to dietary shifts at all but was coopted for diet after its appearance (Martin 1983). Developmentally the increase in thickness during the early to middle Miocene appears to be the result of a lengthened period of enamel accretion, which may reflect a grade shift in general body growth between thin-enamelled ancestral hominoids and later thick-enamelled species (Martin 1983). Such a fundamental reorganization is unlikely to have been the result of dietary pressures alone, for its ramifications for systems other than the dentition are considerable (Martin 1983).

Thick enamel appears to have been retained in the ancestors of both the orang utan and the hominid clades. After the branching of the orang utan clade, the limited fossil evidence available indicates retention of the thick enamel and robust jaw adaptations in the ancestors of the African ape and human clade. The early australopithecines maintained this ancestral pattern, with the inference of no dietary change from ape ancestors, despite the evidence for increasingly seasonal habitats associated with hominid fossils. This conclusion is supported by microwear studies (Grine 1986), and similar conclusions emerge from other lines of inquiry, for example the retained adaptations for ape-like arboreality in the postcrania of *A. afarensis* (Stern & Susman 1983) and the short developmental period of tooth crown formation in australopithecines, again a retained ape characteristic (Beynon & Dean 1988).

The lack of morphological change in the teeth and jaws of the earliest hominids suggests that the dietary heritage from thick-enamelled apes was sufficient for dietary needs during the early stages of human evolution. Subsequently, divergence within the australopithecines occurred, with further change leading to the genus *Homo*, and the manufacture and use of tools both made possible the expansion of human diet into areas not possible for our ape ancestors and brought about the reduction of certain physical attributes no longer needed for food preparation, for example, reduction in tooth size and robusticity and bodily strength. It is reasonable to ask, however, if the earliest hominids used or made tools, and if so, were they different in these respects from their ape ancestors.

Chimpanzees have a great breadth of intelligence and culture, and their use of tools parallels that of early hominids (McGrew 1987; Wynn & McGrew 1989). These authors demonstrated high levels of similarity between chimpanzee technology and the early hominid Oldowan industry, with tool use and manufacture function-related to particular tasks in both chimpanzees and early hominids. Both select appropriate raw materials for making tools, and they make the tools opportunistically without a pre-conceived plan other than to produce a shape or cutting edge that will perform the task in hand. Two differences between apes and the makers of the Oldowan industry, however, are that apes do not use stone for tool manufacture, and they do not use tools to make tools. It is also possible that the hominids used tools in a more sophisticated way, for instance in meat processing (Potts 1988) and even more significantly for obtaining meat during hunting (Wynn & McGrew 1989), with subsequent changes in food processing and diet.

The use of space, in the sense of food preparation areas, is also similar in chimpanzees and early hominids (Potts 1988). Potts shows that the artifact accumulations of early hominids at Olduvai do not represent home bases or workshops but were areas to which both animal remains and raw material for stone tools were transported. He compares this with the transport of nuts and hammer stones to suitable anvils by chimpanzees, or alternatively an anvil and hammer stones carried to the source of nuts (references in Potts (1988)). Both require forward planning, and the only difference from the archaeological situation is the distance the objects are carried, up to 500 m in the case of the chimpanzees and 2–10 km for the hominids.

Because of these characteristics shared by hominids and their nearest ape relatives, it is very likely that they were also present in the common ancestor of apes and humans in the Miocene. There is no preserved cultural record from this far back in time, but neither is there any cultural record from the time of the earliest hominids. It would be surprising, however, if *Australopithecus afarensis* had not achieved at least the degree of tool use and manufacture of chimpanzees, and there is every reason to expect that the common ancestor of humans and chimpanzees had done likewise. The material used, however, would have been biodegradable tissues rather than stone, and to detect their presence we must look to indications from behaviour and morphology of the fossil hominoids.

There are a number of indications from such factors as body size, brain size and morphology, diet, manipulative ability and foraging strategy to provide some insights into the potential for tool use in early hominoids. Larger species are normally dominant over smaller species, have greater control over their environment, are more mobile, and have longer periods of gestation and infancy. As a result, family units may persist longer, and there is a general increase in social behaviour. Larger species also have absolutely larger brains and longer periods of maternal dependence and life spans during which to acquire learned behaviour. Being large also entails an increase in the absolute amount of food needed and a larger supplying area to provide it, and the latter is related to an increase in comparative brain size (CBS (Clutton–Brock & Harvey 1980)). Home range size is also related to the dispersal of resources within it, being necessarily larger when resources are more widely dispersed (Macdonald 1983).

Fruit is a discontinuous resource, both in space and in time (Milton 1988), and being more widely dispersed in the environment than are leaves or grass, frugivores (and predators) tend to have larger home ranges and therefore larger brains (relative to body size) than do folivores and grazers (Clutton–Brock & Harvey 1980). Fruit is also a predictable resource in that fruiting occurs regularly from one year to the next, and this leads to an emphasis on learning and memory together with greater flexibility of behaviour (Milton 1988). It is proposed by Milton, therefore, that larger brains, and by implication greater intelligence, arose initially as a dietary response rather than to facilitate social relations. Many of the thick-enamelled hominoids were large-bodied frugivores close to the ancestry of the great ape and human lineages, and it is likely that they had relatively large brain sizes. Brain size has been calculated for *Proconsul* as 167 cm^3 with a CBS of 48.8 % (Walker *et al.* 1983): this is comparable to chimpanzees and to australopithecines (Aiello & Dean 1990) but higher than gorillas and monkeys.

Proconsul also provides evidence of possibly high manipulative ability, having a long and opposable thumb. Variety of grip was probably greater than is present today in tool-using and tool-making chimpanzees. A detailed comparison with the hand of *A. afarensis* would be of interest, for there appear to be considerable functional similarities. In other respects also, australopithecines show little or no change from fossil apes in brain size, dental development, type of diet, overall body size or even, perhaps, in cultural development and tool use, and as a result it may be concluded that they were the same evolutionary grade as the Miocene (and recent) apes. Subsequent brain enlargement, dental reduction and changes in the gut proportions in the genus *Homo* (Milton 1988) suggest a higher quality diet and food processing, which may be the result of increasingly sophisticated use of tools.

We are grateful to Leslie Aiello, Libby Andrews, John Fleagle, Kaye Reed, Jack Stern, Andy Whiten and Bernard Wood for comments on the manuscript. We are indebted to Roshna Wunderlich for assistance with all stages of the photographic work, and to Robert Kruszynski for drafting the figures. The study of enamel thickness in hominoids would not have been possible without the cooperation and encouragement of those entrusted with the curatorship of the materials used. We are grateful to the staff of the Department of Zoology and the Sub-Department of Anthropology at the British Museum of Natural History for access to the modern material and the *Proconsul*, *Heliopithecus*, *Oreopithecus* and *Sivapithecus* material. We thank the late Dr Crusafont–Pairo for access to the *Dryopithecus* specimen, Professor Louis de Bonis for the *Graecopithecus* specimen, Professor Heinz Tobien for access to the Paşalar specimens. We reiterate the gratitude expressed by Grine & Martin (1988) to F. C. Howell for the *P. boisei* specimens, C. K. Brain for the *P. crassidens* and *P. robustus* specimens, and P. V. Tobias and A. R. Hughes for the *A. africanus* material. The AMR 1810 SEM was made available by the State University of New York at Stony Brook, and the enamel thickness portion of this research is based upon work supported by the National Science Foundation under award No. BNS 89 18695.

6. REFERENCES

Aiello, L. C. 1981 Locomotion in the Miocene Hominoidea. In *Aspects of human evolution* (ed C. B. Stringer), pp. 63–97. London: Taylor and Francis.

Aiello, L. C. & Dean, C. 1990 *Human evolutionary anatomy*. London: Academic Press.

Alcock, B. R. 1984 Robusticity of the mandibular corpus. MSc thesis, University of London.

Alpagut, B., Andrews, P. & Martin, L. 1990 New hominoid specimens from the middle Miocene site at Paşalar, Turkey. *J. hum. Evol.* **19**, 397–422.

Andrews, P. 1978 A revision of the Miocene Hominoidea of East Africa. *Bull. Br. Mus. nat. Hist.* A **30**, 85–224.

Andrews, P. 1981 Species diversity and diet in monkeys and apes during the Miocene. *In Aspects of human evolution* (ed. C. B. Stringer), pp. 25–61. London: Taylor and Francis.

Andrews, P. 1983 The natural history of *Sivapithecus*. In *New interpretations of ape and human ancestry* (ed. R. L. Ciochon & R. Corrucini), pp. 441–463. New York: Plenum Press.

Andrews, P. 1985 Family group systematics and evolution among catarrhine primates. In *Ancestors: the hard evidence* (ed. E. Delson), pp. 14–22. New York: Alan R. Liss.

Andrews, P. 1989 Palaeoecology of Laetoli. *J. hum. Evol.* **18**, 173–181.

Andrews, P. 1990 Palaeoecology of the Miocene fauna from Paşalar, Turkey. *J. hum. Evol.* **19**, 569–582.

Andrews, P. & Cronin, J. 1982 The relationships of *Sivapithecus* and *Ramapithecus* and the evolution of the orang utan. *Nature, Lond.* **297**, 541–546.

Andrews, P., Lord, J. & Evans, E. M. N. 1979 Patterns of ecological diversity in fossil and modern mammalian faunas. *Biol. J. Linn. Soc.* **11**, 177–205.

Andrews, P. & Martin, L. 1987*a* Cladistic relationships of extant and fossil hominoids. *J. hum. Evol.* **16**, 101–118.

Andrews, P. & Martin, L. 1987*b* The phyletic position of the Ad Dabtiyah hominoid. *Bull. Br. Mus. nat. Hist.* A **41**, 383–393.

Beard, K. C., Teaford, M. F. & Walker, A. C. 1986 New wrist bones of *Proconsul africanus* and *P. nyanzae* from Rusinga Island, Kenya. *Folia primat.* **47**, 97–118.

Beynon, A. D. & Wood, B. A. 1986 Variations in enamel thickness and structure in East African hominids. *Am. J. Phys. Anthrop.* **70**, 177–193.

Beynon, A. D. & Dean, M. C. 1988 Distinct dental development patterns in early fossil hominids. *Nature, Lond.* **335**, 509–514.

de Bonis, L., Bouvrain, G., Geraads, D. & Koufos, G. 1990 New hominid skull material from the late Miocene of Macedonia in northern Greece. *Nature, Lond.* **345**, 712–714.

Clutton-Brock, T. H. & Harvey, P. H. 1980 Primates, brains and ecology. *J. Zool.* **190**, 309–323.

Collins, D. A. & McGrew, W. C. 1988 Habitats from three groups of chimpanzees (*Pan troglodytes*) in western Tanzania compared. *J. hum. Evol.* **17**, 553–574.

Fleagle, J. 1978 Size distributions of living and fossil primate faunas. *Paleobiology* **4**, 67–76.

Ghiglieri, M. P. 1987 Sociobiology of the great apes and the hominid ancestor. *J. hum. Evol.* **16**, 319–357.

Gordon, K. D. 1982 A study of microwear on chimpanzee molars: implications for dental microwear analysis. *Am. J. phys. Anthrop.* **59**, 195–215.

Grine, F. E. 1986 Dental evidence for dietary differences in *Australopithecus* and *Paranthropus*: a quantitative analysis of permanent molar microwear. *J. hum. Evol.* **15**, 783–822.

Grine, F. E. & Martin, L. 1988 Enamel thickness and development in *Australopithecus* and *Paranthropus*. In *Evolutionary history of the 'Robust' Australopithecines* (ed. F. E. Grine), pp. 3–42. New York: Aldine de Gruyter.

Harrison, T. 1986 New fossil anthropoids from the middle Miocene of East Africa and their bearing on the origin of the Oreopithecidae. *Am. J. phys. Anthrop.* **71**, 265–284.

Harrison, T. 1988 A taxonomic revision of the small catarrhine primates from the early Miocene of East Africa. *Folia primatol.* **50**, 59–108.

Hartman, S. E. 1988 A cladistic analysis of hominoid molars. *J. hum. Evol.* **17**, 489–502.

Hill, A. & Ward, S. 1988 Origin of the Hominidae: the record of African large hominoid evolution between 14 My and 4 My, *Yearb. phys. Anthrop.* **31**, 49–83.

Hooker, J. J. 1991 British mammalian palaeocommunities across the Eocene–Oligocene transition and their environmental implications. In *Eocene–Oligocene climate and biotic evolution* (ed. D. R. Prothero). Princeton University Press.

Jolly, C. J. 1970 The seed-eaters: a new model of hominid differentiation based on a baboon analogy. *Man* **5**, 5–28.

Kay, R. F. 1977 Diet of early Miocene African hominoids. *Nature, Lond.* **268**, 628–630.

Kay, R. F. 1981 The nut-crackers – a new theory of the adaptations of the Ramapithecinae. *Am. J. phys. Anthrop.* **55**, 141–151.

Kay, R. F. 1984 On the use of anatomical features to infer foraging behaviour in extinct primates In *Adaptations for foraging in non-human primates.* (ed. P. S. Rodman & J. G. H. Cant), pp. 21–53. New York: Columbia University Press.

Kay, R. F. 1985 Dental evidence for the diet of australopithecines. *A. Rev. Anthropol.* **14**, 315–341.

Kay, R. F. & Grine, F. E. 1988 Tooth morphology, wear and diet in *Australopithecus* and *Paranthropus* from southern Africa. In *The evolutionary history of the robust australopithecines* (ed. F. E. Grine), pp. 427–447. New York: Aldine de Gruyter.

Kay, R. F. & Simons, E. L. 1980 The ecology of Oligocene African Anthropoidea. *Int. J. Primatol.* **1**, 21–37.

Leakey, M. D. & Harris, J. M. 1987 *Laetoli – A Pliocene site in Northern Tanzania.* Oxford: Clarendon Press.

Leakey, R. E. & Leakey, M. G. 1986 A new Miocene hominoid from Kenya. *Nature, Lond.* **324**, 143–145.

Lucas, P. W. & Luke, D. A. 1984 Chewing it over: basic principles of food breakdown. In *Food acquisitions and processing in primates* (ed. D. J. Chivers, B. A. Wood & A. Bilsborough), pp. 283–301. New York: Plenum Press.

Macdonald, D. W. 1983 The ecology of carnivore social behaviour. *Nature, Lond.* **301**, 379–384.

Maier, W. 1984 Tooth morphology and dietary specialization. In *Food acquisition and processing in primates* (ed. D. J. Chivers, B. A. Wood & A. Bilsborough), pp. 303–330. New York: Plenum Press.

Martin, L. 1983 The relationships of the later Miocene Hominoidea. PhD thesis, University of London.

Martin, L. 1985 Significant of enamel thickness in hominoid evolution. *Nature, Lond.* **314**, 260–263.

Martin, L. 1986 Relationships among extant and extinct great apes and humans. In *Major topics in primate and human evolution* (ed. B. Wood, L. Martin & P. Andrews), pp. 161–187. Cambridge University Press.

Martin, L. & Andrews, P. 1984 The phyletic position of *Graecopithecus freybergi* Koenigswald. *Cour. Forsch. Inst. Senckenberg* **69**, 25–40.

Marzke, W. M. 1983 Joint function and grips of the *Australopithecus afarensis* hand, with special reference to the region of the capitate. *J. hum. Evol.* **12**, 197–211.

McGrew, W. C. 1987 Tools to get food: the subsistants of Tasmanian aborigines and Tanzanian chimpanzees compared. *J. anthrop. Res.* **43**, 247–258.

Milton, K. 1988 Foraging behaviour and the evolution of primate intelligence. In *Machiavellian intelligence* (ed. R. W. Byrne & A. Whiten), pp. 285–305. Oxford: Clarendon Press.

Napier, J. R. & Davis, P. R. 1959 The forelimb skeleton and associated remains of *Proconsul africanus. Fossil Mammals Afr.* **16**, 1–69.

Potts, R. 1988 *Early hominid activities at Olduvai.* New York: Aldine de Gruyter.

Rose, M. D. 1983 Miocene hominoid postcranial morphology: monkey-like, ape-like, neither, or both? In *New interpretations of ape and human ancestry* (ed. R. L. Ciochon & R. S. Corruccini), pp. 405–417. New York: Plenum.

Rose, M. D. 1989 New postcranial specimens of catarrhines from the middle Miocene Chinji Formation, Pakistan. *J. hum. Evol.* **18**, 131–162.

Ruff, C. B., Walker, A. & Teaford, M. F. 1989 Body mass, sexual dimorphism and femoral proportions of *Proconsul* from Rusinga and Mfwangano Islands, Kenya. *J. hum. Evol.* **18**, 515–536.

Ryan, A. S. & Johanson, D. C. 1989 Anterior dental microwear in *Australopithecus afarensis*: comparison with human and non-human primates. *J. hum. Evol.* **18**, 235–268.

Stern, J. T. & Susman, R. L. 1983 The locomotor anatomy of *Australopithecus afarensis. Am. J. phys. Anthrop.* **60**, 279–317.

Teaford, M. F. 1988 A review of dental microwear and diet in modern mammals. *Scanning Microsc.* **2**, 1149–1166.

Teaford, M. F. & Walker, A. C. 1984 Quantitative differences in dental microwear between primate species with

different diets and a comment on the presumed diet of *Sivapithecus*. *Am. J. phys. Anthrop.* **64**, 191–200.
Walker, A. C. 1981 Diets and teeth. *Phil. Trans. R. Soc. Lond.* B **292**, 57–64.
Walker, A. C., Falk, D., Smith, R. & Pickford, M. 1983 The skull of *Proconsul africanus*: reconstruction and cranial capacity. *Nature, Lond.* **305**, 525–527.
Walker, A. C., Hoeck, H. N. & Perez, L. 1978 Microwear of mammalian teeth as an indicator of diet. *Science, Wash.* **201**, 908–910.
Walker, A. C. & Teaford, M. F. 1989 The hunt for *Proconsul*. *Scient. Am.* **260**, 76–82.
Ward, S. C. & Hill, A. 1987 Pliocene hominid mandible from Tabarin, Baringo, Kenya. *Am. J. phys. Anthrop.* **72**, 21–37.
Ward, S. C. & Pilbeam, D. R. 1983 Maxillofacial morphology of Miocene hominoids from Africa and Indo-Pakistan. In *New interpretations of ape and human ancestry* (ed. R. L. Ciochon & R. S. Corruccini), pp. 211–238. New York: Plenum.
White, T. D. 1977 New fossil hominids from Laetoli, Tanzania. *Am. J. phys. Anthrop.* **46**, 197–230.
Wrangham, R. W. 1980 An ecological model of female-bonded primate groups. *Behaviour* **75**, 262–299.
Wynn, T. & McGrew, W. C. 1989 An apes view of the Oldowan. *Man* **24**, 383–398.

Discussion

L. Aiello (*University College, London, U.K.*). Do the australopithecines have any hominid adaptations other than bipedalism, and are they in fact just bipedal apes?

P. Andrews. Traditionally, hominids have been defined on the basis of three characters: large brains, changes in the teeth, and bipedal locomotion. Recent work has shown that australopithecines are no more encephalized than are some of the apes, for example chimpanzees, and one school of thought also claims that brain structure and complexity are also similar, although there is still some dispute over this. It is shown here that the functional adaptation of the main battery of chewing teeth is essentially unchanged in the early australopithecines from their ape ancestors, although the later australopithecines show some modifications to this pattern, such as the enlargement of the molars and the development of two-cusped third premolars. The incisors of most australopithecines also show no reduction from the ancestral ape condition. We are left, therefore, with bipedalism as the only definitive hominid adaptation, although even here recent work has shown that early australopithecines retain such ape features as shortened hind-limbs and long curved phalanges. It is very likely that the actual appearance of the early australopithecines was ape-like, their manner of life, social grouping and diet also being ape-like, and even their level of intelligence and use of tools was no greater than that of apes, so that the only thing that distinguished the earliest hominids from contemporary species of the ape was their ability to walk or run on two feet instead of four.

A. E. Scandrett (*184 Granby Court, Denbigh, Milton Keynes, U.K.*). Does the enamel thickness of incisors and canines correlate with that of the molars?

P. Andrews. The authors have not sectioned any incisors or canines, and so have no information on how enamel thickness correlates with that of the molars. It would appear unlikely that there is a strong correlation, as the function of the anterior teeth is different from molar function.

A. E. Scandrett. Which teeth were studied to provide data on molar enamel thickness?

P. Andrews. Sample sizes for the three lower molars and the three upper molars were as follows: gorilla, 18 teeth sectioned buccolingually, 6 teeth sectioned mesiodistally; chimpanzee, 16 teeth sectioned buccolingually, 5 teeth sectioned mesiodistally; orang utan, 18 teeth sectioned buccolingually, 6 teeth sectioned mesiodistally; humans, 17 teeth sectioned buccolingually, 7 teeth sectioned mesiodistally; fossil hominoids, 17 teeth sectioned buccolingually. For any one tooth for any one species, the usual sample size was four specimens, but for the upper and lower third molars of chimpanzees the sample sizes were two and three respectively.

Hominid carnivory and foraging strategies, and the socio-economic function of early archaeological sites

ROBERT J. BLUMENSCHINE

Department of Anthropology, Rutgers University, New Brunswick, New Jersey 08903, U.S.A.

SUMMARY

New evidence for the tissue types exploited by early hominids from carcasses possibly acquired through scavenging is derived from the larger mammal bone assemblages from FLK I, level 22 (*Zinjanthropus* floor), and FLKN levels 1 and 2 from Bed I, Olduvai Gorge, Tanzania. Published skeletal part profiles from the two archaeological sites are evaluated using (i) modern observations on the sequence by which carnivores consume carcass parts in order to assess the timing of hominid access to carcasses, and (ii) measurements of flesh and marrow yields to assess the tissue types sought and acquired. These results suggest that the maximization of marrow (fat) yields, not flesh (protein) yields, was the criterion shaping decisions about carcass processing. Because of evidence for density-dependent destruction of some flesh-bearing parts by scavengers of the hominid-butchered assemblages, however, it is uncertain whether carcass parts were transported and acquired by hominids in a largely defleshed condition. The results on tissue types acquired are combined with a discussion of predation risk, feeding competition and the equipment needs of carcass processing in an attempt to identify archaeological test implications of competing hypotheses for the socio-economic function of the earliest archaeological sites.

1. INTRODUCTION

A Stone Age archaeologist may envy the ethologist who can observe diet and foraging behaviour through binoculars rather than by examining the types and locations of food residues and processing tools their subjects leave behind. The earliest, two-million-year-old archaeological evidence for hominid diet and foraging comes from a handful of sites containing flaked stone artifacts and fragmented bones of diverse vertebrate species. Decoding these fossilized traces of behaviour must precede investigation of issues in evolutionary ecology that can engross one studying living primates from the start. The bones at these sites have passed through many taphonomic filters from the time an animal died until its excavated remains are interpreted. These physical and biological filters, whose identification and joint effects are imperfectly understood, introduce many potential observational biases that conspire to distort, but hopefully not mask, the dietary and foraging signals contained in the bone remains.

None the less, the animal-food diet and foraging strategies of early hominids are central issues in palaeoanthropology. The quality of evidence for them is better than for most aspects of prehistoric adaptation. And, the opening lament aside, archaeological pursuit of this issue is important because comparative primate studies, although vital for defining primitive and uniquely derived components of human diet and foraging, cannot replace the fossil record for revealing the exact courses and circumstances of human subsistence evolution (Isaac 1968). The earliest archaeological record is most pertinent because it testifies to the emergence of a unique modification to generalized primate omnivory that involved stone-tool-mediated consumption of larger mammals, acquired possibly through corporate rather than individualistic foraging (Isaac 1978).

Isaac (e.g. 1978, 1980, 1981) was the first archaeologist to systematically explore the relationship between hominid diet and foraging behaviour. His goal was to provide fossil evidence for his home base – food sharing model of the socio-economic function of Plio-Pleistocene sites from Olduvai Gorge, Tanzania, and Koobi Fora, Kenya. Later relabelled as central place foraging (Isaac 1983), Isaac's socio-economic model maintained that meat and marrow from larger mammals, whether acquired by hunting or scavenging, were obtained in a corporate manner and in sufficiently large packets to sustain a system of food reciprocity. Accordingly, early archaeological sites represent places from which hominids practised radiative foraging on a daily basis: plant-gathering and animal-food-acquiring sub-groups procured food in surplus quantities and returned it to the site for intentional or *de facto* sharing among group members. Among other evidence, Isaac used the presence of stone tool cut marks on meat-bearing limb bones (see Bunn *et al.* 1980) to argue that meat from larger mammals was the central commodity of the food sharing system.

Binford (1981, 1986, 1988) countered Isaac's and later Bunn's (1986) position by arguing that hominids scavenged from the nutritionally marginal, mostly marrow, leftovers of carnivore prey. Although Bunn & Kroll (1986, 1988) have shown that the skeletal part data upon which Binford's interpretation was based produced flawed profiles, Binford (1985) maintained that hominids were engaged in minimal food transport and no sharing because carcass yields were too meagre to satisfy even an individual's hunger. Foraging was akin to the 'individualistic-feed-as-you-go' foraging (Isaac, 1983) of non-human primates, a strategy that can be referred to as routed foraging (Binford, 1984).

Others have added to the debate. Shipman (1983, 1986) argued from stone-tool cut mark data for an absence of carcass disarticulation by hominids (but see Bunn & Blumenschine (1986); Gifford-Gonzalez (1989); Lyman (1987)). This indicated to her 'an opportunistic foraging scavenging mode of life' (Shipman 1983) lacking food sharing and other characteristics of human hunting–gathering. Potts (1984) argued that hominids transported carcass parts, acquired through hunting and scavenging (Potts 1983), to caches of stone for butchery. The caches had been pre-established throughout the foraging range in anticipation of eventually finding a carcass nearby. Potts added that danger from carnivores drawn to a cache would render the sites servicable only as briefly visited carcass processing locales, and not home bases. On the basis of a study of modern scavenging opportunities from ungulates in the Serengeti region of Tanzania, Blumenschine (1986*a*, 1987) identified abandoned lion kills as potentially the most regular scavenging opportunity encountered by early hominids. Abandoned lion kills in riparian woodlands are predictably located and typically provide little flesh but mostly marrow and head contents; this yield was suggested to be too low either to finance transport or to sustain an active system of food sharing.

The crux of the debate between Issac and Binford on the socio-economic function of early archaeological sites is not whether hominids hunted or scavenged, but rather whether the nutritional yield (i.e. tissue types and amounts acquired) and regularity of carcass acquisition by hominids was adequate to promote food transport and sharing. Indeed, a range of potential scavenging opportunities for early hominids other than that from lion kills have been identified by Blumenschine (1986*a*, 1987), by Cavallo (Cavallo & Blumenschine 1989) from tree-stored leopard kills, and by Marean (1989; see also Blumenschine (1987)) from saber-tooth cat kills. These studies show that scavenging may have been a high-yield, low-risk, predictable source of animal foods for hominids.

Here, I attempt to make explicit linkages between carcass acquisition and foraging by hominids. The links are defined by various motivations and constraints on carcass part transport from a death site. These in turn serve to define the socio-economic function of early archaeological sites. One major link, as argued above, is the nutrient yield and availability of carcass parts acquired. I use direct determinations of marrow and flesh masses and naturalistic observations on carcass consumption by carnivores to interpret skeletal part data from the two largest and best reported Plio-Pleistocene archaeological bone assemblages. These sites are FLKI level 22 (the '*Zinjanthropus* floor'), and FLKN levels 1 and 2. Both date between 1.8 and 1.7 Ma from Bed I of Olduvai Gorge, Tanzania, and were deposited in fine-grained sediments near the shore of a shallow lake that occupied the Olduvai Basin (Leakey 1971). FLK Zinj contains 60000 plus bones, approximately 3350 of which can be identified to derive from a minimum of 48 larger mammal individuals (Bunn & Kroll 1986). FLKN 1/2 has 45 such individuals represented by 2274 identified bone specimens (Bunn 1986). Most larger mammal bones at both sites derive from species of the family Bovidae.

2. NUTRIENT YIELD AND CARCASS CONSUMPTION SEQUENCES

Skeletal part profiles have been used by archaeologists to address modes of carcass acquisition by hominids. Blumenschine (1986*b*) argued that the carcass consumption sequence, i.e. the order in which carcass parts are depleted of flesh and within-bone tissues (marrow, brain, etc.) by modern larger mammalian and avian carnivores, is the most appropriate model available for interpreting skeletal part profiles to these ends. So far it and alternative models (Binford

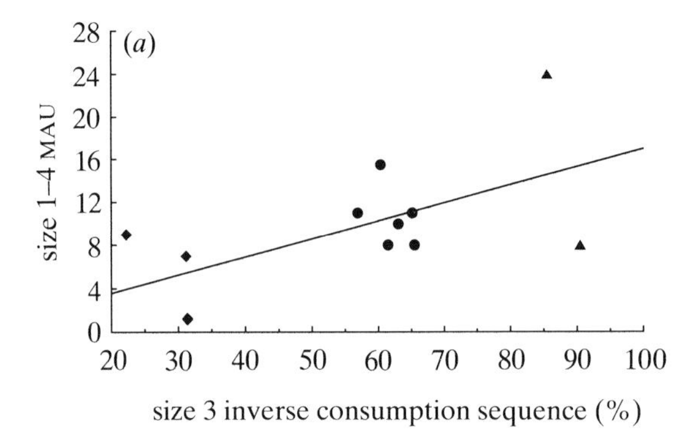

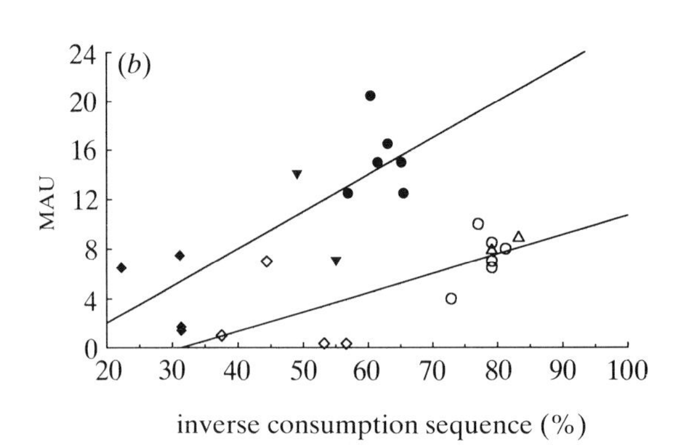

Figure 1. Carcass part availability and MAU representation at FLK Zinj (*a*) and FLKN 1/2 (*b*). Inverse consumption sequence values are based on Blumenschine (1986), and MAU values are from Bunn (1986) (see table 1). Open symbols, size 1 and 2 mammals; closed symbols, size 3 and 4 mammals. Diamonds, flesh from post-cranial axial parts; inverse triangles, flesh from head parts; circles, marrow from long bones; upright triangles, contents of head parts. Regression statistics: (*a*) $y = 0.17x+0.24$, $r = 0.61$, $p = 0.04$; (*b*) size 1 and 2 mammals, $y = 0.16x-4.9$, $r = 0.72$, $p = 0.008$; size 3 and 4 mammals, $y = 0.30x-3.9$, $r = 0.80$, $p = 0.002$.

Table 1. *Raw data used for analysis of skeletal part profiles*

(Skeletal parts used are those provided by Bunn (1986) for FLK Zinj. and FLKN 1/2. S12 = size 1 and 2; S34 = size 3 and 4 (see footnote a for definition of animal size groups).)

skeletal part	yield/g marrow[a]	yield/g flesh[b]	inverse consumption sequence[e] marrow	inverse consumption sequence[e] flesh	bulk density[g]	FLK ZINJ MAU[i]	FLKN 1/2 MAU[i]
head							
cranium					0.61[h]	8	16
S12	146.8[c]	303.3	83.8	45.8	—	—	9
S34	325[c]	1065.5	90.5	55.1	—	—	7
mandible					0.61	24	22
S12	10[d]	303.3	79.2	54.2	—	—	8
S34	10[d]	1065.5	85.6	49.1	—	—	14
post-cranial axial							
vertebrae					0.30	1.3	2.07
S12	—	6543.4	—	56.6[f]	—	—	0.33
S34	—	11507.3	—	31.4[f]	—	—	1.74
ribs					0.40	1.19	1.77
S12	—	7054.1	—	53.2	—	—	0.35
S34	—	18752.7	—	31.4	—	—	1.42
pelvis					0.49	9	7.5
S12	—	1596.0	—	37.5	—	—	1
S34	—	5185.4	—	22.3	—	—	6.5
scapula					0.49	7	14.5
S12	—	3000.4	—	44.4	—	—	7
S34	—	8239.8	—	31.2	—	—	7.5
long bones							
humerus					0.63	10	25
S12	19.96	1723.6	79.2	47.8	—	3	8.5
S34	48.08	4830.2	63.1	29.3	—	7	16.5
radius-ulna					0.68	11	22
S12	18.32	766.1	79.2	53.2	—	3	7
S34	67.26	1704.8	65.2	45.2	—	8	15
femur					0.57	11	16.5
S12	32.52	9384.2	72.9	33.3	—	4	4
S34	53.66	16053.5	56.9	22.2	—	7	12.5
tibia					0.74	15.5	30.5
S12	46.52	1340.6	77.1	41.7	—	6	10
S34	117.5	2557.2	60.4	36.6	—	9.5	20.5
metacarpal					0.72	8	19
S12	14.3	—	79.2	—	—	3	6.5
S34	27.7	—	65.5	—	—	5	12.5
metatarsal					0.74	8	23
S12	16.66	—	81.3	—	—	5	8
S34	26.34	—	61.5	—	—	3	15

[a] Long bone values are average adult wet masses of marrow in diaphyseal cavities (Blumenschine & Madrigal, in preparation). Size 1–2 based on two Thomson's gazelle, three Grant's gazelle, and two impala; size 3–4 based on five wildebeest. Whole-animal values are given (i.e. single bone yield is multiplied by 2 for paired parts) in keeping with use of MAU values for archaeological sites.

[b] Based on average adult male wet masses from Blumenschine & Caro (1986). Size 1–2 based on percent of total flesh mass for two Thomson's gazelle, two Grant's gazelle and two impala multiplied by the average total flesh mass (31919 g) for the Grant's gazelle and impala. Size 3–4 based on three wildebeest. Whole-animal values are given (i.e. single bone yield is multiplied by 2 for paired parts). Blumenschine & Caro's head flesh values are equally divided between cranium and mandible, whereas flesh yields for vertebrae are the sum of their neck flesh and lumbar flesh. Rib flesh includes thoracic vertebrae flesh.

[c] Based on brain mass (325 g) of one adult male wildebeest (size 3) equalling 0.46% of total flesh mass. Size 1–2 brain mass therefore = 31919 (total flesh mass) × 0.0046.

[d] Based on one subadult zebra (R. J. Blumenschine, unpublished data).

[e] Average percentage of parts encountered upon discovery of a carcass that remained with some flesh, marrow or other contents (see Blumenschine 1986). Values for size 3–4 based on consumption sequence for size 3 carcasses only.

[f] Average values for cervical vertebrae, lumbar vertebrae and rib cage.

[g] Based on Lyman's (1984) photondensitometer measurements for densest part of each bone sampled.

[h] Based on Lyman's (1984) value for the mandible, as cranium has similar density as the mandible in other studies cited by Lyman.

[i] From Bunn (1986, tables 2 and 4). Size-specific MAU are only available for long bones at FLK Zinj.

Table 2. *Correlation and regression statistics relating the inverse consumption sequence, yield of flesh and marrow parts, and bulk density to* MAU *values from FLK Zinj and FLKN 1/2*

(Values are Pearson's correlation coefficient (top), probability (middle), and the slope ± s.e. for the least squares regression line (bottom). The latter is not provided for bulk density. S12, size 1 and 2; S34, size 3 and 4; S14, size 1 to 4.)

	FLK Zinj.			FLKN 1/2		
	S12	S34	S14[a]	S12	S34	S14
inverse consumption sequence						
model IA						
post-cranial axial flesh,	—	—	0.61	0.71	0.56	—
head contents, long bone	—	—	0.04	0.008	0.06	—
marrow ($n = 12$ parts)	—	—	0.17 ± 0.07	0.16 ± 0.05	0.17 ± 0.07	—
model IB						
post-cranial axial flesh,	—	—	0.39	0.43	0.80	—
head flesh, long bone	—	—	0.21	0.16	0.002	—
marrow ($n = 12$ parts)	—	--	0.15 ± 0.11	0.09 ± 0.06	0.30 ± 0.07	—
model II						
flesh values only	—	—	0.38	−0.08	0.17	—
($n = 10$ parts)	—	—	0.28	0.83	0.64	—
	—	—	0.23 ± 0.20	−0.04 ± 0.17	0.10 ± 0.20	—
log yield (grams)						
long bone marrow	0.68	0.96	0.91	0.18	0.71	0.76
($n = 6$ or 12)	0.14	0.003	0.0001	0.73	0.11	0.004
	4.3 ± 2.4	8.9 ± 1.4	7.3 ± 1.0	1.9 ± 5.1	8.6 ± 4.3	13.2 ± 3.6
flesh, all parts	—	—	−0.67	−0.67	−0.49	—
($n = 10$)	—	—	0.04	0.03	0.15	—
	—	—	−95 ± 3.8	−4.7 ± 1.8	−6.8 ± 4.2	—
flesh, long bones	0.07	−0.63	0.23	−0.73	−0.61	0.03
($n = 4$ or 8)	0.93	0.37	0.59	0.27	0.40	0.94
	0.2 ± 2.1	−1.8 ± 1.5	1.2 ± 2.1	−4.0 ± 2.7	−4.7 ± 4.4	0.4 ± 4.9
bulk density						
all parts ($n = 12$)	—	—	0.56	0.82	0.88	—
	—	—	0.06	0.001	0.0001	—
long bones ($n = 6$)	0.44	−0.21	—	0.68	0.42	—
	0.38	0.68	—	0.14	0.40	—

[a] Uses size 3 values for inverse consumption sequence and log yield except for analyses restricted to long bones, where size-specific models are used.

1984; Potts 1983) have been applied to only a handful of Pleistocene bone assemblages.

The carcass consumption sequence is very regular for a wide range of carnivores (lions, cheetah, spotted hyena, vultures) and larger mammal carcass sizes (Blumenschine 1986*a*, *b*). In its regional expression the sequence proceeds from (i) hindquarter flesh to (ii) forequarter flesh, (iii) head flesh, (iv) hindlimb marrow, (v) forelimb marrow, and (vi) head contents (brain, and pulps of the mandible and cranium) (see also Haynes (1982)).

The regular sequence of carcass consumption has an energetic basis. The consumption sequence values for parts from medium-sized ungulates are significantly correlated with flesh mass. This suggests that carnivores are maximizing their rate of feeding by preferentially eating from skeletal units that offer the highest tissue yields (Blumenschine 1986*b*). The fundamental basis of the consumption sequence justifies the uniformitarian assumption that prehistoric mammalian carnivores consumed carcass parts in a similar order, and that the consumption sequence can be used to interpret skeletal part profiles at archaeological sites with regard to the timing of hominid access to carcasses. Specifically, if hominids had first access to a carcass (as in hunting), they should have acquired an anatomically even distribution of skeletal parts, or one skewed toward the highest yielding parts that are consumed early in the sequence. Late access to partially consumed carcasses (scavenging), on the other hand, should be characterized by the availability of food from parts consumed only in the latter part of the consumption sequence. The consumption sequence can be expressed in units that reflect the percentage of time I encountered a carcass part with some edible tissue remaining on or in it (i.e. the inverse consumption sequence). Hominid hunting is modelled by a negative relationship between a bone assemblage's skeletal part representation and the inverse consumption sequence, and scavenging by a positive one (Blumenschine 1986*b*; see, for example, figure 1).

3. YIELD AND AVAILABILITY OF SKELETAL PARTS AT FLK Zinj. AND FLKN 1/2

The abundance of different skeletal parts from FLK Zinj and FLKN 1/2 is expressed in Minimum Animal Units (MAU; Binford 1984), as determined by Bunn

(1986; Bunn & Kroll 1986). These values (table 1) are an expression of the minimum number of animals required to account for the skeletal parts present. If all individuals are represented by complete skeletons, the MAU values for each skeletal unit would be equal†. MAU values for all skeletal units are available for size 1–2 and size 3–4 mammals‡ for FLKN 1/2; published data on FLK Zinj permit such size discrimination only for long bone MAUs.

Skeletal part proportions at the two sites are uneven with respect to those in whole animals. The paucity of post-cranial axial parts, particularly ribs and vertebrae, may indicate the selective transport of long bones and heads from the death site to the archaeological site (Bunn 1986; Bunn & Kroll 1986). Alternatively, Marean *et al.* (1991) have shown quantitatively that the pattern is consistent with the preferential deletion of the uncommon parts by site scavengers after hominid butchery occurred. Here, I use the carcass consumption sequence to investigate whether the uneven skeletal part profiles might also be related to the differential availability of carcass foods at a death site.

Three consumption sequence models are used (table 2). Each is based on the alternative availability of edible tissues for those parts that bear flesh and contain either marrow (long bones) or other edible tissues (cranium and mandible). Models IA and IB reflect the standard availability of edible tissues from abandoned lion and cheetah kills (Blumenschine 1987), which, if not thoroughly defleshed (model IA), will retain flesh lastly on the head (model IB). Model II uses values for flesh only.

The best fit consumption sequence models for both sites (i.e. those with the highest Pearson's correlation coefficient) are models IA and IB, where significantly positive relationships are found; model II yields poorly correlated, statistically insignificant results (figure 1*a*, *b*, table 2). These results, particularly the significantly positive slopes of the regression equations, are consistent with scavenging of carcasses that had been largely defleshed before hominid access. The greater abundance at both sites of long bones and head parts compared to post-cranial axial flesh-bearing parts is therefore consistent with the high probability (large inverse consumption sequence value) of finding abandoned predator kills with defleshed long bones that still contain marrow, and head parts, which, if not remaining with flesh, still contain the brain and pulps.

The above indications of scavenging are supported by data on the flesh and marrow yields of bones at the two sites. Both assemblages show strong, significantly positive relationships between marrow yield and long bone MAUs (figure 2, table 2; Blumenschine & Madrigal, in preparation). These relationships are made more intriguing by the fact that small, size 1–2 long bones are in general less abundant that those of large mammals in a manner predicted by their lower marrow yield: for FLK Zinj, the least squares regression line describing the relationship for size 1–2 bones is very similar to that for size 3–4 bones. Further, the slopes of the size 1–4 regression lines are similar for both sites.

In contrast to marrow yields, flesh yields are negatively, and, with the exception of size 3–4 animals from FLKN 1/2, significantly related to skeletal part abundance (figure 3*a*, *b*, table 2). Further, when long bone flesh yields are considered alone, there are no significant relationships within or between size classes, unlike those found for marrow yields (table 2). The results on tissue yields suggest that skeletal part representation at the two sites is related to marrow extraction and not flesh consumption, as is consistent with hominid scavenging.

These indications of scavenging, however, may be partly spurious. The low representation of post-cranial axial flesh-bearing parts might not result from their largely defleshed state and consequent neglect by hominids at the death site, but rather from their preferential destruction subsequent to defleshing by hominids. This alternative is supported by the significant, or nearly significant, positive relationships at both sites between MAU and bulk density of bones (table 2). Post-cranial axial bones have the lowest bulk density (table 1), making them less resistant to destructive agents than denser long bones and head parts.

Density-dependent skeletal part profiles have been suggested to have many potential causes (Lyman 1984). The most likely causal agent at the sites in question, however, is ravaging of hominid-butchered bone by site scavengers such as hyenas. Carnivore coprolites and tooth-marked bone are present at both sites (Bunn 1986; Leakey 1971). Also, recent experiments demonstrate a strong positive relationship between bone density and the survivorship of bones fed on by spotted hyenas (Marean & Spencer 1991). Indeed, the number of vertebrae, pelves, ribs, and hindlimb bones at the two sites under examination are significantly and positively correlated to their survivorship under the influence of spotted hyena ravaging (Marean *et al.* 1991).

The representation of long bones, however, is not density-dependent. No consistent and significant relationships are obtained between long bone MAUs from either site and their bulk density (table 2). This is due to the fact that Bunn's long bone MAUs are based not only on articular ends, but also midshaft fragments. Bunn has argued that midshafts must be used to estimate long bone abundance accurately. This assertion has been confirmed by experiments showing that hyenas selectively remove grease-filled long bone epiphyses from hammerstone-broken assemblages of long bones, but virtually ignore hammerstone-generated midshafts which have been deprived of any nutrient value (Blumenschine 1988; Binford *et al.* 1988; Marean & Spencer 1991; Marean *et al.* 1991).

† An MAU value of 1 for humeri would be obtained if two humeri were present (even if both were from the left side), because an animal has two humeri. Likewise, the presence of 27 vertebrae would yield an MAU based on vertebrae of 1.

‡ Larger mammal size classes used here are zooarchaeological standards following Bunn *et al.* (1980). Size 1 animals are less than 50 lb. live mass (e.g. Thomson's gazelle), size 2 are 50–250 lb. (e.g. Grant's gazelle), size 3 are 250–750 lb. (e.g. wildebeest and zebra), and size four are 750–2500 lb. (e.g. buffalo). 1 lb. = 0.4536 kg.

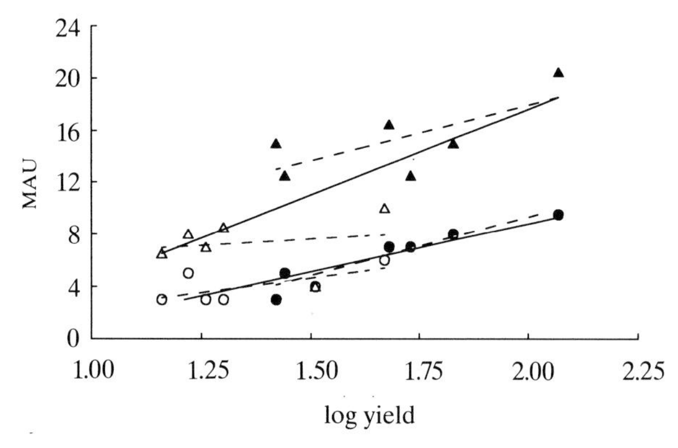

Figure 2. Marrow yields and MAU representation of long bones at FLK Zinj (circles) and FLKN 1/2 (triangles). Marrow yields (in g) are from Blumenschine & Madrigal (in preparation) and MAU values are from Bunn (1986) (see table 1). Dashed least squares regression lines are shown for size 1 and 2 bones (open symbols) and size 3 and 4 bones (closed symbols). Regression statistics (solid lines) for the combined size groups at FLK Zinj: $y = 7.3x - 5.9$, $r = 0.91$, $p = 0.0001$; and FLKN 1/2: $y = 13.2x - 8.8$, $r = 0.76$, $p = 0.004$.

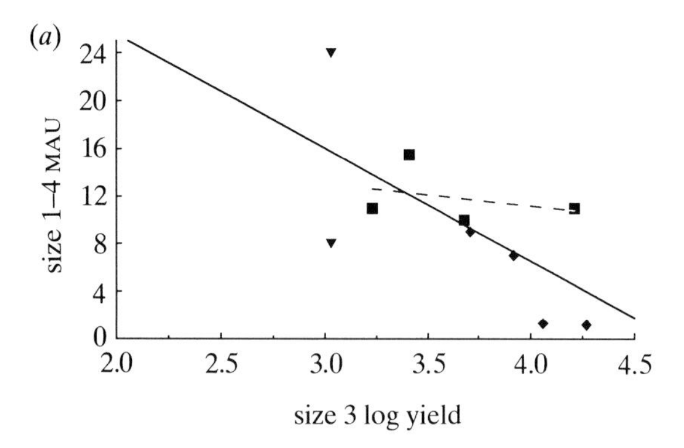

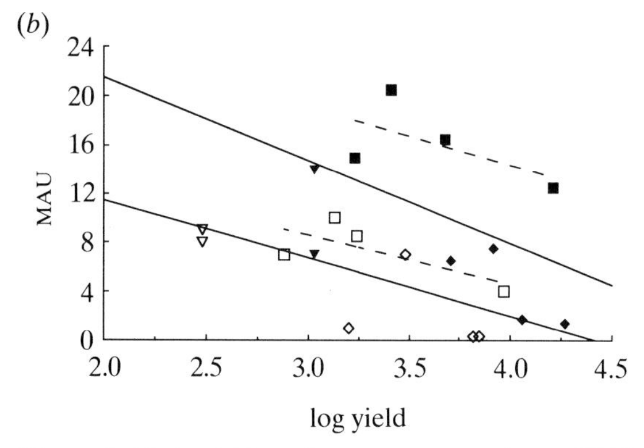

Figure 3. Flesh yields and MAU representation at FLK Zinj (*a*) and FLKN 1/2 (*b*). Flesh yields are based on Blumenschine & Caro (1986), and MAU values are from Bunn (1986) (see table 1). Open symbols, size 1 and 2 mammals; closed symbols, size 3 and 4 mammals. Diamonds, post-cranial axial bones; inverse triangles, head bones; squares, long bones. Dashed least squares regression lines are for long bones only and are based on only four points per size class as metapodials bear no flesh. Regression statistics (solid lines) are for all parts: (*a*) $y = -9.5x + 44.6$, $r = -0.67$, $p = 0.04$; (*b*) size 1 and 2 mammals, $y = -4.7x + 20.9$, $r = -0.67$, $p = 0.03$; size 3 and 4 mammals, $y = -6.8x + 35.2$, $r = -0.49$, $p = 0.15$.

4. DIET AND FORAGING

The results are consistent with hominid scavenging of mostly marrow and head contents, but are rendered equivocal owing to evidence for density-dependent destruction of some flesh-bearing parts. Ambiguity in interpretations can be reduced by focussing on long bones, which do not show density-dependent patterning, and by segregating discussion according to three component behaviours of animal food foraging by hominids. Listed in order of increasing equivocality, these are (i) marrow bone breakage, (ii) the tissue types transported from death sites, and (iii) the tissue types encountered at death sites.

Minimally, the results presented for both sites show that long bones were broken in direct proportion to their marrow yield. That hominids rather than carnivores were responsible for extracting marrow from most long bones is indicated by new data from FLK Zinj (Blumenschine & Bunn 1990) on the proportion of long bone fragments that bear percussion marks produced by hammerstone breakage (Blumenschine & Selvaggio 1988). Hence, although both assemblages contain disproportionate numbers of 'meaty' upper limb bones at the expense of 'non-meaty' metapodials (Bunn 1986; Bunn & Kroll 1986), the results show that long bone representation is conditioned by marrow (= fat) yield. Hominid breakage of marrow bones at the two sites followed an energy maximizing strategy, as might be expected if hominid populations were energy limited. Because flesh yields correlate negatively with carcass part abundance, hominids do not seem to have been protein limited.

Higher-level inferences are more equivocal. Long bones might have been transported to the sites in proportions determined by marrow yield. This would suggest that most parts were transported in a defleshed condition and that the motivation for bone transport was not flesh-sharing. Alternatively, other long bones may have been transported in proportions predicted by maximization of flesh yield, but only those which were broken for their marrow survived post-butchery ravaging because they are represented mainly by unnutritious midshaft fragments. Or, the results could indicate that flesh from many axial parts and long bones was butchered at the death site and transported without the bone, as among some hunter-gatherers today (see, for example, Bunn *et al.* (1988)).

Even more equivocal is the interpretation based on the consumption sequence analysis that hominids typically encountered most bones in a defleshed condition. This interpretation apparently contradicts Bunn's (Bunn *et al.* 1980; Bunn & Kroll 1986) inference that the presence of stone tool cut-marking on 'meaty' long bones implies hominid access to fully-fleshed or largely intact carcasses, as in early-access scavenging or hunting. However, there are no experimental or ethnographic models that permit one to equate the proportion of parts that bear cut marks with the proportion of bones that were actually defleshed or disarticulated. Further, we do not know how to distinguish cut-marking produced while defleshing whole muscle masses from that inflicted during the removal of scraps of flesh that commonly survive carnivore consumption (Blumenschine in Bunn & Kroll (1986)). Likewise, for carnivore tooth marks on bones from the two sites, we need to learn the distinction between tooth-marking associated with defleshing and that produced during bone breakage for marrow or grease extraction. An integrated analysis of the anatomical distribution of stone tool cut marks, hammerstone percussion marks, and carnivore tooth marks is needed to assess the edible tissues consumed by carnivores versus hominids, and the order in which they fed on carcasses. Such an analysis for long bones from FLK Zinj is in progress (Blumenschine & Bunn 1990). Until this independent evidence is available, however, the skeletal part data suggests that the presence at the sites of long bones that once bore meat requires neither hominid transport of long bone flesh, nor the availability of fully-fleshed bones at the death site. Skeletal part transport may instead have been based upon maximization of energy yields from marrow fat, and hominids may have typically encountered most bones already defleshed by carnivores.

5. THE SOCIO-ECONOMIC FUNCTION OF EARLY ARCHAEOLOGICAL SITES

The nutrient benefit of animal food foraging is not the only factor relevant for evaluating carcass part transport decisions made at the death site and the socio-economic function of the archaeological site to which parts were transported. These factors are defined in figure 4*a*, *b* by four sets of circumstances, including (i) predation risk, (ii) interspecific and intraspecific feeding competition, (iii) the availability of carcass processing equipment at the death site, and (iv) food yield encountered and food yield per individual in a transport party.

Carcass part transport may be constrained by energy yields and by logistical concerns (see O'Connell *et al.* (1988) for an ethnographic example). Routed foraging, involving minimal or no transport, should occur if a carcass' energy yield is too low to finance transport, if adequate processing equipment is available at the death site, and if competition and predation risk at the carcass is low (figure 4*a*). However, if transport is energetically and logistically feasible, several motivations for transport that serve to define the socio-economic function of the transport site have been specified. These can be evaluated in terms of the nutritional, logistical and ecological circumstances prevailing at the death site and transport site.

In Isaac's home base model, sharing intentions in an atmosphere of intragroup cooperation was the explicitly hypothesized motivation for the transport of surplus quantities of animal foods to archaeological sites. High carcass yields are shown in figure 4*a* to be necessary to finance transport to a central place, which, on average, will be further from the death site than the nearest refuge tree or source of stone for carcass processing. Field butchery of flesh to reduce transport costs to a central place is shown to be constained by predation risk: high risk would allow only the quick quartering of carcasses into transportable packets. A site to which carcass parts were transported can serve as a home base only if each transporter, on average, provides a surplus, and if predation risk and inter- and intraspecific feeding competition is low (figure 4*b*).

Inadequate amounts of carcass processing equipment (stone tools) at the death site are shown in figure 4*a* to promote transport to a nearby cache of usable stone created previously by hominids. Caching is not indicated in the context of low predation risk, as Potts feels that hominid involvement with carcasses would always expose them to danger and competition from carnivores. Hence, if archaeological sites were stone caches of the type envisioned by Potts, they were the setting for only hasty processing and consumption, but no other maintenance and social activities. The availability of surplus quantities of food at the cache is relevant only in terms of the amount that was wasted (figure 4*b*).

High risk of predation or interspecific competition from carnivores at the death site would promote transport to a locality where consumption can occur in relative safety (Isaac 1983). A foraging mode with this motivation can be labelled refuging. Refuging is unmotivated by equipment needs and should occur virtually regardless of available yield (figure 4*a*). If early archaeological sites were refuges they should be characterized by low levels of predation risk and competition, but, unlike central places, places to which food surpluses were not introduced intentionally (figure 4*b*). Dissemination of food among group members may occur: low available yields might render this akin to what Isaac characterized as the 'tolerated scrounging' seen during meat-eating among chimpanzees. Occasional high yields might result in '*de facto* sharing' (Isaac 1983).

Other socio-economically relevant motivations for carcass part transport may have operated. For instance, intraspecific competition at the death site might encourage individual hominids to remove carcass parts to an area where feeding could occur in isolation. Such behaviour typifies the strategy of spotted hyenas when feeding at carcasses in large groups.

Archaeologically visible criteria for measuring these constraints and motivations on carcass part transport are needed if the model in figure 4 is to serve as more than an heuristic devise for comparing competing hypotheses on hominid foraging and socio-ecology. This was the goal of the skeletal part analysis reported here with respect to carcass part yield. Although those results were equivocal in and of themselves, it should also be noted that archaeologists, including Isaac and Binford, have not been able to test whether animal foods were ever present at the sites in amounts required to sustain active sharing. They are hindered by a lack of direct evidence for (i) the number of hominids occupying these sites at any one time, (ii) the duration over which the bone assemblages accumulated (*contra* Potts 1986; see Bunn & Kroll 1987; Lyman & Fox 1989), (iii) the relative contribution of plant foods to hominid subsistence (Sept 1986), and (iv) an ignorance of whether the archaeologically visible activities of

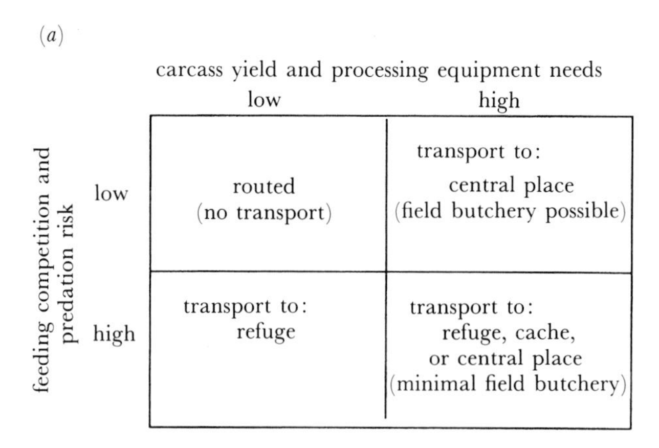

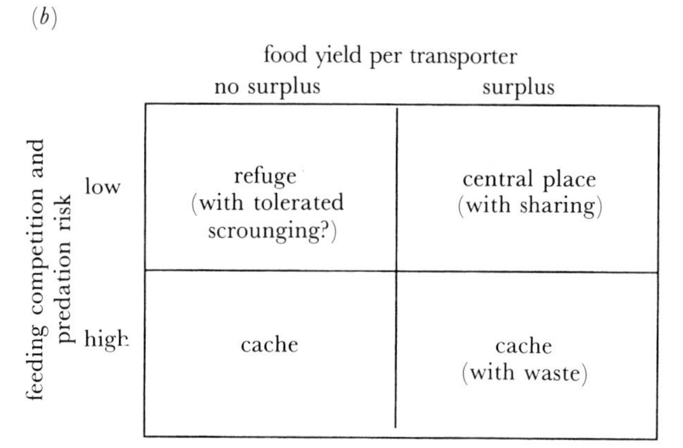

Figure 4. (*a*) Nutritional, logistical and ecological circumstances at the death site that define the type of site to which carcasses are transported, and (*b*) circumstances at the transport site that define the socio-economic function of the site.

stone tool discard and carcass processing were typical or rare elements of hominid adaptation. None the less, if long bone marrow was the main benefit of animal food foraging, the food value introduced to the sites seems small. Based on data in table 1, the long bones at FLK Zinj and FLKN 1/2 contained a total of 3234 and 6743 grams of marrow, respectively. Assuming that (i) all marrow was composed soley of fat, (ii) there are 9 Kcal g^{-1} of fat, and (iii) that the minimum daily caloric requirement of an adult Plio-Pleistocene hominid was 2000 Kcal, then the marrow bones at FLK Zinj represent 14.6 person-days of caloric intake, and those at FLKN 1/2 represent 30.3 person-days. Assuming further that one carcass was acquired each day the sites were used by a hominid group, the minimum number of mammal carcasses at FLK Zinj (48) and FLKN 1/2 (45) indicate a daily availability of 30% and 67% of a single adult's daily caloric requirements, respectively. This energy yield may have been adequate to provision infants (Isaac 1983; C. Peters, personal communication), but rather than promoting cooperative sharing, such small amounts of easily defended, energy-dense marrow might instead encourage intragroup competition. Clearly, these estimates of animal food value at the sites should be seen as minima. Nonetheless, if animal-food sharing was anywhere as regular and economically central as it is in modern hunter-gatherers, the cut mark data will have to show that flesh was the mainstay of the system.

Archaeological detection of the degree of predation risk and competition with carnivores is also complicated. Potts (1984) argued for a high degree of competition and risk on the basis of the presence of carnivore bones at the sites, and evidence for hasty carcass processing from the presence of some complete, unbroken long bones. However, given the tree-climbing abilities of Plio-Pleistocene hominids (Susman & Stern 1982), carcass parts should have been transported to the vicinity of climable trees where ready refuge from predators could be found if necessary. Such refuging, as in savanna-woodland baboons today, should characterize hominid foraging even if they were not carnivorous. Any additional risk encountered at animal death sites could be minimized by scavenging from abandoned kills (Blumenschine 1987), as is consistent with results of the consumption sequence analysis presented here.

Evidence for the adequacy of processing equipment at the death site will require experimental studies that examine the minimum amount of stone needed to process a carcass of a particular size and completeness. If this amount is small, then such equipment may have been routinely carried by hominids on foraging expeditions. If larger, as I suspect it was for complete processing of most larger mammal carcasses, sites might represent a combination of refuges and *de facto* stone caches. That is, sites are places originally chosen for carcass processing on the basis of the refuge they offered; the same place would be revisited with locally acquired carcass foods because usable stone remaining from prior processing episodes could supplement that which was routinely carried. This site function can accommodate the interpretation of skeletal part profiles presented here, and could account for the accumulation of the bone and stone at FLKN 1/2 and FLK Zinj.

The socio-economic function of sites such as FLK Zinj and FLKN 1/2 is in these ways conditioned by opportunities and constraints on hominid diet and foraging. Their actual socio-economic function will become apparent only if we gain a better understanding of (i) the timing of carnivore and hominid involvement with carcass parts and the tissues consumed by each, (ii) the amount and types of stone tools needed to butcher a carcass of a particular size and completeness, and (iii) the ecological setting of the sites, including the proximity of refuge trees and the extent of competition between hominids and carnivores. Most of these issues are currently the subject of investigation among Early Stone Age archaeologists, the results of which will lead to a greater understanding of the nature and circumstances of hominid subsistence and social evolution.

I am grateful to Sal Capaldo, Curtis Marean and Andrew Whiten for comments on the manuscript.

REFERENCES

Binford, L. R. 1981 *Bones: ancient men and modern myths*. New York: Academic Press.
Binford, L. R. 1984 *Faunal remains from Klasies River Mouth*. New York: Academic Press.

Binford, L. R. 1985 Human ancestors: changing views of their behavior. *J. anthrop. Archaeol.* **4**, 292–327.

Binford, L. R. 1986 Comment on Bunn and Kroll's 'Systematic butchery by Plio-Pleistocene hominids at Olduvai Gorge'. *Curr. Anthropol.* **27**, 444–446.

Binford, L. R. 1988 Fact and fiction about the *Zinjanthropus* floor: data, arguments and interpretations. *Curr. Anthropol.* **29**, 123–135.

Binford, L. R., Mills, M. G. L. & Stone, N. M. 1988 Hyena scavenging behavior and its implications for the interpretation of faunal assemblages from FLK 22 (the Zinj floor) at Olduvai Gorge. *J. anthrop. Archaeol.* **7**, 99–135.

Blumenschine, R. J. 1986*a* *Early hominid scavenging opportunities: implications of carcass availability in the Serengeti and Ngorongoro ecosystems.* Oxford: British Archaeological Reports International Series 283.

Blumenschine, R. J. 1986*b* Carcass consumption sequences and the archaeological distinction of scavenging and hunting. *J. hum. Evol.* **15**, 639–659.

Blumenschine, R. J. 1987 Characteristics of an early hominid scavenging niche. *Curr. Anthropol.* **28**, 383–407.

Blumenschine, R. J. 1988 An experimental model of the timing of hominid and carnivore influence on archaeological bone assemblages. *J. archaeol. Sci.* **15**, 483–502.

Blumenschine, R. J. & Bunn, H. T. 1990 A preliminary report on surface modifications to long bones from FLK *Zinjanthropus*, Olduvai Gorge, Tanzania (Research report submitted to the L. S. B. Leakey Foundation, Oakland, California, April 1990).

Blumenschine, R. J. & Caro, T. M. 1986 Unit flesh weights of some East African bovids. *J. Afr. Ecol.* **24**, 273–286.

Blumenschine, R. J. & Madrigal, T. C. Long bone marrow yields of some East African ungulates. (In preparation.)

Blumenschine, R. J. & Selvaggio, M. M. 1988 Percussion marks on bone surfaces as a new diagnostic of hominid behaviour. *Nature, Lond.* **333**, 763–765.

Bunn, H. T. 1986 Patterns of skeletal representation and hominid subsistence activities at Olduvai Gorge, Tanzania and Koobi Fora, Kenya. *J. hum. Evol.* **15**, 673–690.

Bunn, H. T., Bartram, L. E. & Kroll, E. M. 1988 Variability in bone assemblage formation from Hadza hunting, scavenging, and carcass processing. *J anthrop. Archaeol.* **7**, 412–457.

Bunn, H. T. & Blumenschine, R. J. 1987 On "theoretical framework and tests" of early hominid meat and marrow acquisition: A reply to Shipman. *Am. Anthropol.* **89**, 444–448.

Bunn, H. T., Harris, J. W. K., Isaac, G., Kaufulu, Z., Kroll, E., Schick, K., Toth, N. & Behrensmeyer, A. K. 1980 FxJj 50: an Early Pleistocene site in northern Kenya. *World Archaeol.* **12**, 109–136.

Bunn, H. T. & Kroll, E. M. 1986 Systematic butchery by Plio/Pleistocene hominids at Olduvai Gorge, Tanzania. *Curr. Anthropol.* **27**, 431–452.

Bunn, H. T. & Kroll, E. M. 1987 On butchery by Olduvai hominids – a reply to Potts. *Curr. Anthropol.* **28**, 96–98.

Bunn, H. T. & Kroll, E. M. 1988 Reply to Fact and fiction about the *Zinjanthropus* floor. *Curr. Anthropol.* **29**, 135–149.

Cavallo, J. A. & Blumenschine, R. J. 1989 Tree-stored leopard kills: expanding the hominid scavening niche. *J. hum. Evol.* **18**, 393–399.

Gifford-Gonzalez, D. 1989 Shipman's shaky foundations. *Am. Anthropol.* **91**, 180–186.

Haynes, G. 1982 Utilization and skeletal disturbances of North American prey carcasses. *Arctic* **35**, 266–281.

Isaac, G. Ll. 1968 Traces of Pleistocene hunters: an East African example. In *Man the hunter* (ed. R. B. Lee & I. DeVore), pp. 253–261. Chicago: Aldine.

Isaac, G. Ll. 1978 Food-sharing and human evolution: archaeological evidence from the Plio-Pleistocene of East Africa. *J. anthropol. Res.* **34**, 311–325.

Isaac, G. Ll. 1980 Casting the net wide: a review of archaeological evidence for early hominid land use and ecological relations. In *Current argument on early man* (ed. L.-K. Konigsson), pp. 226–251. Oxford: Pergamon for the Swedish Academy of Sciences.

Isaac, G. Ll. 1981 Stone Age visiting cards: approaches to the study of early land-use patterns. In *Pattern of the past: studies in the honour of David Clarke* (ed. I. Hodder, G. Ll. Isaac & N. Hammond), pp. 131–155. Cambridge University Press.

Isaac, G. Ll. 1983 Bones in contention: competing explanations for the juxtaposition of early Pleistocene artifacts and faunal remains. In *Animals and archaeology*, vol. 1 (Hunters and their prey (ed. J. Clutton-Brock & C. Grigson), pp. 3–19. Oxford: British Archaeological Reports International Series 163.

Leakey, M. D. 1971 *Olduvai Gorge*, vol. 3 (Excavations in Beds I and II, 1960–1963). Cambridge University Press.

Lyman, R. L. 1984 Bone density and differential survivorship of fossil classes. *J. anthropol. Archaeol.* **3**, 259–299.

Lyman, R. L. 1987 Hunting for evidence of Plio-Pleistocene hominid scavengers. *Am. Anthropol.* **89**, 710–715.

Lyman, R. L. & Fox, G. L. 1989 A critical evaluation of bone weathering as an indication of bone assemblage formation. *J. archaeol. Sci.* **16**, 293–317.

Marean, C. W. 1989 Sabertooth cats and their relevance for early hominid diet and evolution. *J. hum. Evol.* **18**, 559–582.

Marean, C. W. & Spencer, L. M. 1991 Impact of carnivore ravaging on zooarchaeological measures of element abundance. *Am. Antiquity.* (In the press.)

Marean, C. W., Spencer, L. M., Blumenschine, R. J. & Capaldo, S. D. 1991 Captive spotted hyena bone choice and destruction, the Schlepp effect, and Olduvai archaeofaunas. *J. archaeol. Sci.* (In the press.)

O'Connell, J. F., Hawkes, K. & Blurton-Jones, N. 1988 Hadza hunting, butchery, and bone transport and their archaeological implications. *J. anthropol. Res.* **44**, 113–161.

Potts, R. B. 1983 Foraging for faunal resources by early hominids at Olduvai Gorge, Tanzania. In *Animals and archaeology*, vol. 1 (Hunters and their prey) (ed. J. Clutton-Brock & C. Grigson), pp. 51–62. Oxford: British Archaeological Reports International Series 163.

Potts, R. B. 1984 Home bases and early hominids. *Am. Sci.* **72**, 338–347.

Potts, R. B. 1986 Temporal span of bone accumulations at Olduvai Gorge and implications for early hominid foraging behavior. *Paleobiology* **12**, 25–31.

Sept, J. M. 1986 Plant foods and early hominids at site FxJj 50, Koobi Fora, Kenya. *J. hum. Evol.* **15**, 751–770.

Shipman, P. 1983 Early hominid lifestyles: hunting and gathering or foraging and scavenging? In *Animals and archaeology*, vol. 1 (Hunters and their prey) (ed. J. Clutton-Brock & C. Grigson), pp. 31–49. Oxford: British Archaeological Reports International Series 163.

Shipman, P. 1986 Scavenging or hunting in early hominids: Theoretical framework and tests. *Am. Anthropol.* **88**, 27–43.

Susman, R. L. & Stern, J. T. 1982 Functional morphology of *Homo habilis*. *Science, Wash.* **217**, 931–934.

Discussion

A. Whiten (*Scottish Primate Research Group, University of St Andrews, U.K.*). My question is about the extent to which scavenging would have been a drastic tactic for a primate to adopt. On the occasions I have observed baboons come across a lion kill they have given it a wide berth, despite the fact that when they make their own kills the meat is highly valued and intensely competed over. In Dr Blumenschine's carcass watches, has he observed primates' reactions?

R. J. Blumenschine. My observations of over 250 fresh carcasses in the Serengeti include only one episode in which non-human primates (olive baboons and vervet monkeys) influenced the consumption of a scavengeable carcass. A large troop of baboons displaced a female leopard with her large cub from their resting tree that was adjacent to another harbouring a half-eaten Thomson's gazelle kill (Cavallo & Blumenschine 1989). The baboon troop fed for several hours on pods from the *Acacia* resting tree, never having become obviously aware of the carcass despite its close proximity. Later, John Cavallo (personal communication) observed the same troop driving the same female leopard off another tree-stored kill. This time, two adult baboons inspected the carcass briefly without feeding on it before joining the rest of the troop to feed on pods at the base of the storage tree. For all other, terrestrial carcasses I observed, baboons were simply not in the vicinity, suggesting that their range use is constrained by that of larger predators like lions and spotted hyenas.

Despite anecdotal exceptions (see, for example, Hasegawa *et al.* (1983)) scavenging by non-human primates seems rare (see Strum 1983). It has never been reported on carcasses larger than themselves. Even if they shun the drastic tactic of usurping prey from predators larger and more social than the leopard, non-human primates are simply not equipped to benefit nutritionally from vertebrate foods that come in packages too large to deflesh and disarticulate with their teeth and hands, and whose bones are too stout to be broken (for marrow and brains) manually. This constraint was not present for hominids possessing the simple slicing and pounding tools of the earliest stone industry. Hunting large mammals, given the lack of obvious projectile weaponry for the great majority of prehistory, can be argued to have been a far more difficult tactic for acquiring these, and a less parsimonious explanation for their regular presence at early archaeological sites.

References

Hasegawa, T., Hiraiwa, M., Nishida, T. & Takasaki, H. 1983. New evidence on scavenging behavior in wild chimpanzees. *Curr. Anthrop.* **24**, 231–232.

Strum, S. C. 1983 Baboon cues for eating meat. *J. hum. Evol.* **12**, 327–336.

K. Hawkes (*Department of Anthropology, University of Utah, Salt Lake City, U.S.A.*). Dr Blumenschine briefly mentioned the results of his experiments with Curtis Marean on hyena carcass processing, but did not refer to them in his conclusions. Would he comment further on those results and their implications for interpreting the Oldowan?

R. J. Blumenschine. The work with Curtis Marean (Marean *et al.* 1991) shows that captive spotted hyenas preferentially remove axial parts (pelves, vertebrae and ribs), then limb ends, from assemblages of experimentally butchered bones, leaving behind an assemblage dominated by limb shafts. Preliminary analyses of similar experiments conducted with free-ranging carnivores in the Serengeti by Sal Capaldo support this result. Such selective removal of axial parts can produce skeletal part profiles that mimic those attributed to selective transport of limbs from death sites to archaeological sites by hominids, as argued by Henry Bunn (see Bunn 1986). I identified in my paper a third process that might produce the limb-dominated skeletal part profiles, namely, unselective acquisition by hominids of the only parts that remain with scavengeable food on a carcass defleshed and abandoned by felids. The equifinality can only be broken by using independent lines of evidence, which I suggest in the paper to derive from ongoing studies of the anatomical distribution and incidence of carnivore tooth marks and stone tool butchery marks on bone surfaces.

Because the relevant experiments and archaeological applications are in progress, my conclusions were explicitly equivocal. We do not know whether the bone assemblages were created at death sites, or if they represent whole-carcass transport to a processing locale followed by hyaenid scavenging, unselective transport of parts from carcasses found largely defleshed, or selective transport of limbs from more fully fleshed carcasses acquired by hunting or scavenging. What we have learned from my paper is that marrow bones were broken in direct proportion to their fat yields, a nutritional strategy for which low-risk scavenging of abandoned kills in riparian habitats is totally consistent. Clearly, our analyses and models need further refinement, and promising efforts cited above are in progress.

K. Hawkes. Dr Blumenschine did not mention the work on bone transport patterns among the Hadza presented by Bunn *et al.* (1988), and by our own group (O'Connell *et al.* 1988, 1990). Could he be invited to address the implications of this work for the interpretations of variation in body-part representation in archaeological assemblages, especially with respect to Isaac and Bunn's arguments about the evidence for large body part transport at Plio-Pleistocene sites such as Koobi Fora and Olduvai?

R. J. Blumenschine. O'Connell *et al.* (1988) have shown that the skeletal parts transported by the Hadza from kill sites to camp sites vary among and within prey taxa. Inclusion of Bunn *et al.*'s (1988) data set, argued by Bunn's team to be consistent with the premise that limb-dominated assemblages signal selective food transport by hominids to ancient 'camp' sites, does not change their conclusion: for most taxa, Hadza preferentially transport axial parts, often having processed and discarded limb bones at or near the death site. O'Connell *et al.* (1990) conclude that limb-dominated skeletal-part profiles are not a secure signature of food transport by hominids, a behaviour upon which the veracity of the food sharing or home-base model rests (Isaac 1978; Bunn 1986). This conclusion is strengthened by recent demonstrations that two processes in addition to hominid transport decisions can produce limb dominated assemblages (see my response to Professor Hawkes' first question and my paper in this symposium).

The Hadza results, however, do not as yet provide incisive criteria for interpreting skeletal part profiles with regard to prehistoric hominid transport decisions or site function. Hadza behaviour is archaeologically relevant only to the extent that the mechanisms which transform their behaviour into bony residues are understood and shown to be sufficiently fundamental as to have conditioned prehistoric residues of behaviour in the same way. O'Connell's team has made important strides in explaining variability in the proportion of a carcass' skeleton that is transported from a death site through recourse to fundamental energy optimizing mechanisms related to carcass size, size of transport parties, and distance of a kill site to a camp site. However, as O'Connell's group states, they currently lack the data (net nutrient yields of carcass parts, as these vary with carcass taxon, size and age) which are crucial for explaining variability in the particular bones – for example, axial *versus* limb – that are transported. Until these data are available, archaeologists must rely on criteria other than skeletal part profiles to assess hominid transport of carcass foods and its implications for the socio-economic function of archaeological sites.

Reference

O'Connell, J. K., Hawkes, K. & Blurton-Jones, N. 1990 Reanalysis of large mammal body part transport among the Hadza. *J. archaeol. Sci.* **17**, 301–316.

Ecology and energetics of encephalization in hominid evolution

R. A. FOLEY AND P. C. LEE

Department of Biological Anthropology, University of Cambridge, Downing Sreet, Cambridge CB2 3DZ, U.K.

SUMMARY

Hominid evolution is marked by very significant increase in relative brain size. Because relative brain size has been linked to energetic requirements it is possible to look at the pattern of encephalization as a factor in the evolution of human foraging and dietary strategies. Major expansion of the brain is associated with *Homo* rather than the Hominidae as a whole, and the energetic costs are likely to have forced a prolongation of growth rates and secondary altriciality. It is calculated here that modern human infants have energetic requirements approximately 9% greater than similar size apes due to their large brains. Consideration of energetic costs of brain allow the prediction of growth rates in hominid taxa and an examination of the implications for life-history strategy and foraging behaviour.

1. INTRODUCTION

Human evolution is characterized by two main systematic changes: the adoption of upright, bipedal locomotion and the enlargement and elaboration of the brain. Whereas the first of these has long been thought of as directly related to foraging behaviour, brain expansion has largely been treated outside an ecological framework. The flexible and sophisticated cognition and behaviour resulting from a large brain have been sufficient for explaining the marked rate of increase since the appearance of the hominid clade. From an ecological perspective, however, it is perhaps more interesting to ask what the costs of encephalization are and how they can be afforded. This question is prompted by the simple observation that given the obvious advantages of high encephalization, its occurrence in the living world is relatively rare (Parker 1990). In considering, for example, the development of human foraging strategies, increased returns for foraging effort and food processing may be an important pre-requisite for encephalization, and in turn a large brain is necessary to organize human foraging behaviour.

Two reasons may be put forward to justify the view that the costs of brain growth and maintenance are essential to understanding the conditions under which encephalization may occur. The first derives from allometric studies. Martin (1981, 1983) has shown that mammalian brain size scales allometrically with body size, and that the relation can be described by the regression equation:

$$E = 1.77W^{0.76},$$

where E = brain mass in grams and W = body mass in grams. Several phylogenetically specific equations have also been derived, especially for primates (Martin 1989), and are used where appropriate.

Allometric studies allow brain size to be studied comparatively across species independent of body size (Jerison 1973; Martin 1981, 1983). Encephalization quotients (EQ) represent the positive or negative residual value of brain mass, calculated by:

$$EQ = BM^{o}/BM^{e},$$

where BM^{o} = actual brain mass and BM^{e} = predicted brain mass for body size.

The coefficient of the allometric equation is close to 0.75, similar to the relationship between body size and basal metabolic rate (BMR) (Kleiber 1961). This implies that brain size and BMR are isometrically related, from which the further inference may be drawn that the size of an individual's brain is closely linked to the amount of energy available to sustain it (Milton 1988; Parker 1990). This suggests that whatever selective pressures there may be driving the size of the brain up, these are satisfied only in the context of there being sufficient energy.

The second point proposed by Martin is that the costs of brain growth are borne by the mother. Most brain growth occurs *in utero* and during the post-natal period before weaning (Martin 1983). Brains are metabolically expensive tissue (Passmore & Durnin 1955), and any increase in the size of the brain will act as a considerable drain on maternal energetics. Sustaining high levels of encephalization must therefore be compatible with female, particularly maternal, energetics which acts as a necessary condition for brain enlargement.

The task of this paper, therefore, is to consider the energetic and foraging implications of encephalization in hominid evolution. In particular we shall attempt to quantify the additional energetic costs of larger brains than those of other primate species and to specify where in the course of hominid evolution these may have become significant and therefore required

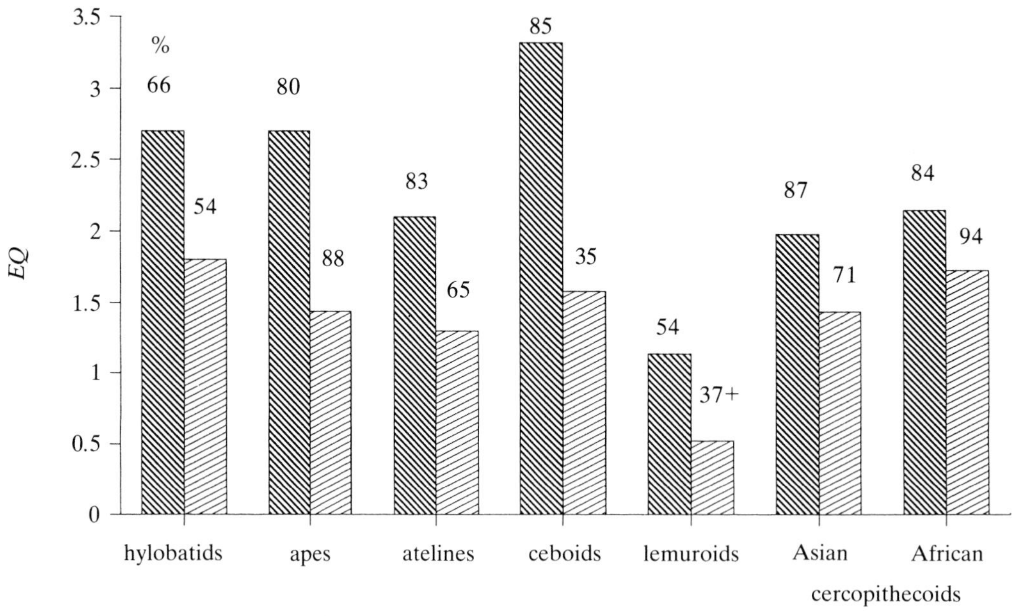

Figure 1. Pairwise comparison of phylogenetically comparable species of primate with the most extreme preference for high-quality (dark-shaded bars) and low-quality (light-shaded bars) foods. Species: Hylobatids: *Hylobates lar* (HQ), *H. symphalangus* (LQ) (Davies *et al.* 1984); African apes: *Pan troglodytes* (HQ), *Gorilla gorilla* (LQ) (Wrangham 1986, Smuts *et al.* 1987); Atelines; *Alouatta palliata* (HQ), *Ateles paniscus* (LQ) (Smuts *et al.* 1987); Cebids: *Cebus capucinus* (HQ), *Callicebus moloch* (LQ) (Smuts *et al.* 1987); Lemuroids: *Lemur catta* (HQ), *Indri indri* (LQ) (Richard 1985); Asian cercopithecoids: *Macaca nemistrina* (HQ), *Presbytis johnii* (LQ) (Davies *et al.* 1984); African cercopithecoids: *Cercocebus albigena* (HQ), *Colobus badius* (LQ) (Smuts *et al.* 1987). HQ = high-quality feeder, LQ = low-quality feeder.

changes in foraging behaviour. To this end we will first examine the relation between primate foraging behaviour and encephalization.

As a caveat it should be stated that the question of the ultimate causation of encephalization is not considered here. It has been argued that the selective advantages of large brains derive from both social (Jolly 1966; Humphrey 1976; Byrne & Whiten 1988) and ecological pressures (Parker & Gibson 1979; Clutton-Brock & Harvey 1980; Gibson 1986). Here we suggest that whatever the cause of encephalization may be, it is predicated upon energetic and hence ecological conditions (Martin 1983; Milton 1988; Foley 1990). If this is the case then social and ecological factors are likely to be closely related, as indeed may be indicated by the increasing evidence for a strong relation between social and reproductive strategies and ecological conditions (Standen & Foley 1989).

2. BRAIN SIZE AND PRIMATE FEEDING BEHAVIOUR

Two major aspects of brain size and complexity have been related to foraging behaviour among the non-human primates. The first of these is the statistical observation that folivores tend to have smaller brains for their body masses than do frugivores (Clutton-Brock & Harvey 1980). The second is that a diet of 'embedded foods' – foods that are neither visually obvious nor available without considerable time and energy costs associated with processing, and hence extracted from a secondary source – require complex cognitive skills (Gibson 1986). The relatively small brains of folivores are therefore thought to derive from the ubiquitous distribution of leaf foods and their relative ease in extraction from the canopy. A further factor that is significant from the point of view of encephalization is that leaves constitute a relatively low level of nutritional quality. Leaf foods generally provide less energy than fruits as much of the potential carbohydrate energy is bound up in structural cellulose (Hladik 1978; Milton 1984). Furthermore such foods are often defended with secondary compounds, which are either toxic or digestion inhibitors (Waterman 1984). A leaf diet among the non-human primates requires the ability to digest cellulose and hemicellulose through microbial fermentation as in the colobines, or bulk feeding as in the gorillas and gelada. Ingestion of toxic or low quality foliage correlates with low basal metabolic rates for body size in many arboreal folivores (McNab 1978) and in strepsirhine primates (Rasmussen & Izard 1988). Taken together, leaf foods require less cognitive skills to procure and ingest, more digestion time within the daily activity cycle, and additional costs in terms of time and energy for their detoxification.

Fruit eating poses different problems. In general fruits supply relatively easily digested energy as well as protein and fatty acids. The location of ripe fruits, enhanced by potential colour differences with background vegetation may not be difficult, but does require spatial and temporal memory of fruit tree locations and renewal rates (Waser & Homewood 1979) when they are widely and patchily distributed. Many 'fruits' are consumed as seeds, which require extraction from tough casings (Gibson 1986). These are linked to perceptual abilities (the recognition that the edible component is within an inedible substrate) and the manual dexterity and coordination to remove such foods. Among most frugivorous primates, fruits are often consumed in combination with other energy- or protein-rich foods such as invertebrates, honey, tree exudates and vertebrates. All these types of food require extractive or capture skills while producing more consumable energy than do leaves. Although the detoxification of unripe fruits can also be a problem (Waterman 1984), frugivores appear to obtain a greater energy intake per gram consumed, even with the increased costs of location and extraction. More complex brain coordination and integration, and memory skills are associated with fruit diets and large brain for body masses.

Treated comparatively the data on quality of diet and primate brain size per unit of body mass shows a positive correlation (Clutton-Brock & Harvey 1980; Gibson 1986; Milton 1988), but there is considerable variance for at least two reasons. The first is that there are major phylogenetic differences in encephalization quotient among the primate grades (Martin 1989), the second is that whereas species of primate with low quality diets seldom have high *EQ*s, there is no certainty that those with high quality diets will have high *EQ*s as their additional energy may be put to other uses.

The relation between foraging behaviour and *EQ* can best be examined in a sample of primates where there is clear dietary preference for either high or low quality foods. A phylogenetically based pairwise examination of the most extreme folivores and frugivores in each of a number of primate clades (figure 1) shows that the folivore in each case has a substantially lower *EQ*. This is supported statistically. When phylogeny at the family and superfamily level is held constant, a regression of *EQ* on percentage of feeding time on leaves, fruits and animals yields a high level of association ($r^2 = 0.74$, $p < 0.01$ for a negative relationship between folivory and *EQ*; $r^2 = 0.686$, $p < 0.05$ for a positive relationship between frugivory and *EQ*; and $r^2 = 0.709$ $p < 0.02$ for a positive relationship between faunivory and *EQ*; $n = 15$ in all cases). Although high *EQ*s are not invariably associated with high quality diets, low quality diets do not appear to permit or be associated with high *EQ*s. Fruit and animal diets appear to be able to fuel larger primate brains, whereas the low energy and high toxicity of folivorous diets may constrain the available energy and basal metabolic rates.

3. PATTERN OF BRAIN SIZE EVOLUTION IN HOMINIDS

The pattern of brain size increase in hominid evolution has been well established from the cranial and endocranial fossil evidence. Fossil hominid cranial capacity varies from 410 cm^3 (KNM-WT17000 (Walker *et al.* 1986)) to 1750 cm^3 in the Amud 1 Neanderthal (Ogawa *et al.* 1970). The earliest taxon, *Australopithecus afarensis*, at approximately 3.0 Ma has a cranial capacity of 450 cm^3, whereas modern humans, with an origin at about 0.1 Ma have an average of approximately 1350 cm^3. This represents a threefold increase in absolute brain volume, or an increase of about 1 cm^3 per 150–200 generations. The rate of increase is however not constant, with the most rapid rise occurring in the last 1.0 Ma.

EQ for fossil hominids can be calculated where body mass can be estimated from post-cranial dimensions (McHenry 1988; Aiello & Dean 1990). Figure 2 shows the relationship between hominid *EQ* and time. From these data it can be seen that encephalization is primarily a characteristic of the genus *Homo* rather than of the hominids as a whole, and that again the increase occurs relatively late. In particular it should be noted that the *EQ* of the earliest hominids is not significantly greater than the upper range observable in living hominoid apes.

From the ecological perspective paramount here, it may be asked whether there is any link between the increase in encephalization and energetic or foraging parameters, supporting the hypothesis that increases in brain size are constrained by ecological factors. At a speculative level it may be noted that evidence for a significant level of both tool use and meat eating is broadly equated with the date of first appearance of the genus *Homo* (Foley 1987*a*, *b*). Although such an interpretation is contested (Binford 1984), it may be argued that tool use increases extractive efficiency or widens the range of resources available (Torrence 1983, Gibson 1986) and that meat represents a high quality resource, both of which would constitute increases in the net rate of energetic returns necessary for further encephalization.

A more specific expectation, given the proposed relationship between encephalization and maternal energetics, is that increasing brain size should be reflected in growth patterns (Shea 1987). It has long been noted (Schultz 1969) that modern humans are characterized by (secondarily) altricial young and delayed maturation in comparison with other primates. Although the contrast between humans and chimpanzees in life-history patterns may not be as marked as has previously been supposed (Lee 1989), the difference is still significant, especially in terms of cognitive development (see Parker 1990). It may be asked whether this difference is related to brain size. Recent work on dental development patterns (Bromage & Dean 1985; Beynon & Wood 1988; Beynon & Dean 1988) has shown that all hominids display the extended maturation pattern of modern humans. Early

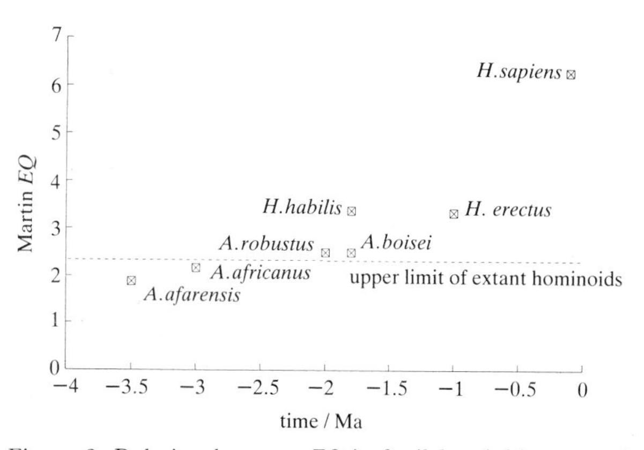

Figure 2. Relation between *EQ* in fossil hominid taxa and time. Data from Aiello & Dean (1990).

Table 1. *Energetic costs of brain maintenance in humans and chimpanzees for the first five years of life*

	human			chimpanzee		
age/ years	brain size	average daily energy requirements kJ	cumulative energy requirements kJ × 10^6	brain size	average daily energy requirements kJ	cumulative energy requirements kJ × 10^6
0	345	364	0.133	172	182	0.066
1	765	807	0.427	277	292	0.173
2	959	1012	0.797	333	351	0.301
3	1058	1116	1.204	362	382	0.441
4	1142	1205	1.644	366	386	0.581
5	1206	1273	2.108	366	386	0.723

(greater than 1.5 Ma) African hominids were dentally more advanced for a given chronological age, based on enamel incremental patterns, than modern humans. Even neanderthals may have matured marginally faster than modern humans (Stringer *et al.* 1990). The pattern of encephalization tracks the slowing down of growth rates in hominids, which in turn indicates changes in female energetics and by implication maternal strategies (Foley 1991).

4. ENERGETIC COSTS OF THE HUMAN BRAIN

Preliminary consideration of the palaeontological evidence for brain enlargement and growth patterns in hominid evolution seems to provide some support for the proposal that the costs of large brains may have been a significant factor in the timing and pattern of encephalization. Before examining in more detail the actual growth patterns of the fossil hominids it is necessary to look at the costs of brain maintenance to determine whether these costs are energetically significant. One reason for doing this is that the data presented so far are equally compatible with an obstetric hypothesis: that is, that because of pelvic constraints on neonate brain size hominids were forced into a more altricial strategy as the adult brain size increased. Although such a hypothesis would not explain why there should be such a prolonged period of post natal growth, it would at least explain why the initial neonate state should be relatively small.

The energetic costs of brain maintenance have been assessed by Passmore & Durnin (1955) as 3.5 ml O_2 per 100 g per minute or 0.07329 kJ per 100 g per minute. This accounts for 20% of human basal oxygen uptake and energy expenditure. Growth costs are additional to this. These basic estimates can be used to calculate the relative costs of brain maintenance of chimpanzees, used here as a baseline for hominid evolution, and modern humans for the first five years of life.

Chimpanzees and modern humans have different brain growth patterns. Chimpanzees are born with a neonate brain mass that is 47% of adult brain mass, and adult size is reached by 4 years of age. In contrast modern humans have neonate brain masses that are 25% of adult mass and by 4 years have reached only 84.1% (Passingham 1982). Table 1 shows the average daily costs involved in brain maintenance for chimpanzees and humans for the first five years of life. Even in its more atricial state, the cost of the human neonate brain exceeds that of the more precocial chimpanzee, and by the age of five years the costs are three times as great: 1273 kJ day^{-1} compared with 386 kJ day^{-1}. When considered cumulatively it can be seen that the chimpanzee brain will have required 0.723 million kJ, at an average of 350 kJ per day, compared with 2.108 million kJ, at an average of about 1000 kJ per day. Overall the energetic costs of brain maintenance for modern humans are about three times those of a chimpanzee. Growth costs will also be commensurately larger. It should be pointed out that both chimpanzee and some human mothers can lactate for five years, and therefore these costs can be partially borne by the mother. The partitioning of energy between that derived from milk or solid foods during the first five years of brain development is of concern primarily in its effect on future maternal reproduction. From a trophic point of view, the overall costs to the mother are greatest during lactation and less during gestation and after the infant begins to feed independently.

Expansion of the brain incurs considerably higher energetic costs whether or not these costs are spread over a longer period of time, with implications for foraging behaviour. To put this into the context of the overall energetic needs of the infants, Ulijaszek & Strickland (1991) have presented data showing that among Gambian infants the average energy requirement in the first eighteen months of life is 2556 kJ per day. Using the estimates of brain maintenance costs provided above it can be deduced that of this total figure 495 kJ per day are required for the brain. Were those infants to have chimpanzee-sized brains we can calculate their reduced energy requirements. The average daily energy requirements of the chimpanzee brain during the first eighteen months of life are 273 kJ. With this size brain, average daily requirements would be:

2556 (total human costs) − 495 (human brain costs) + 273 (chimp brain costs) = 2334 kJ.

This reduction of 223 kJ per day implies that human

Table 2. *Estimates of energy costs of brain maintenance for hominid taxa at different growth rates*

(BM = adult brain mass estimated from fossil specimens (Aiello & Dean 1990; calculated from cranial capacity ($BM = 1.06CC^{0.98}$ (Martin 1989))). The brain mass of the hominids from neonate to age 5 calculated as a percentage of adult brain mass assuming (*a*) chimpanzee growth rates; (*b*) chimpanzee growth rate + 25% shift to human growth rates; (*c*) chimpanzee growth rate + 50% shift to human growth rates; and (*d*) chimpanzee growth rate + 75% shift to human growth rates. Energy costs calculated at 105 kJ per 100 g of brain mass per day (Passmore & Durnin 1955). Figures in bold indicate costs exceeding modern humans at the same age; asterisks indicate costs less than chimpanzees at the same age. At human growth rates all figures for fossil hominids are less than or equal to those of modern humans.)

		age/years					
taxon	*BM*	0	1	2	3	4	5
(*a*)							
A. afarensis	422	208	337	403	439	443	443
A. africanus	454	224	363	434	472	477	477
A. boisei	513	237	383	459	499	504	504
H. habilis (ER 1813)	627	**309**	500	599	952	659	659
H. rudolfensis (ER 1470)	698	**345**	557	667	726	733	733
H. ergaster (ER 3733)	728	**389**	629	753	819	828	828
H. erectus (Sangiran)	857	**423**	**684**	818	891	900	900
H. erectus (Solo)	1057	**522**	**843**	**1010**	**1099**	**1110**	**1110**
H. sapiens (Archaic)	1397	**632**	**1022**	**1223**	**1330**	**1345**	**1345**
Modern humans	1350	290	650	823	905	974	1032
(*b*)							
A. afarensis	422	186	315	381	417	426	430
A. africanus	454	200	339	410	448	458	463
A. boisei	513	212	358	433	474	484	489
H. habilis (ER 1813)	627	277	468	566	619	632	639
H. rudolfensis (ER 1470)	698	308	520	630	689	704	711
H. ergaster (ER 3733)	728	**348**	588	712	778	795	803
H. erectus (Sangiran)	857	**378**	639	774	846	864	873
H. erectus (Solo)	1057	**466**	**788**	**954**	**1043**	**1065**	**1076**
H. sapiens (Archaic)	1397	**565**	**955**	**1157**	**1264**	**1291**	**1304**
Modern humans	1400	290	650	823	905	974	1032
(*c*)							
A. afarensis	422	160*	292	359	390	407	417
A. africanus	454	172*	315	386	419	439	448
A. boisei	513	181	333	408	443	464	474
H. habilis (ER 1813)	627	237	435	533	580	606	619
H. rudolfensis (ER 1470)	698	264	484	594	645	674	689
H. ergaster (ER 3733)	728	**298**	546	670	728	761	778
H. erectus (Sangiran)	857	**324**	594	729	792	828	846
H. erectus (Solo)	1057	**400**	**732**	**977**	**977**	**1021**	**1043**
H. sapiens (Archaic)	1397	**484**	**888**	**1090**	**1184**	**1238**	**1264**
Modern humans	1400	290	650	823	905	974	1032
(*d*)							
A. afarensis	422	137*	275*	337	368	390	408
A. africanus	454	148*	296	363	396	420	439
A. boisei	513	156*	312	383	418	443	464
H. habilis (ER 1813)	627	204	408	500	547	579	606
H. rudolfensis (ER 1470)	698	227	454	557	608	645	674
H. ergaster (ER 3733)	728	257	513	629	687	728	761
H. erectus (Sangiran)	857	279	558	684	747	792	828
H. erectus (Solo)	1057	**344**	**688**	**843**	**921**	**977**	1021
H. sapiens (Archaic)	962	**417**	**834**	**1022**	**1116**	**1184**	**1237**
Modern humans	1350	290	650	823	905	974	1032

infants up to the age of 18 months are approximately 8.7% more energetically costly than chimpanzees due to their high level of encephalization. The foraging implications are that in some way human mothers (or other helpers) must increase the infant's level of nutritional intake compared to the hominoid baseline used here.

Despite the relative crudity of these estimates they demonstrate the basic proposal that encephalization is predicated upon energetic conditions, and therefore that we should expect a close link between expansion of the brain and changes in foraging behaviour and other measures of parental effort, regardless of the precise selective pressures leading to encephalization. As will be discussed below, the energetics of encephalization are a significant factor in human foraging and dietary strategies.

5. ENCEPHALIZATION AND DELAYED MATURATION

It is beyond the scope of this paper to explore the precise foraging behaviour of the early hominids in the context of encephalization. However, because the various hominid taxa of the Pliocene and Pleistocene are characterized by different sized brains it is pertinent to ask where along the continuum from ape sized to human sized brains the costs are likely to become significant. As this is closely tied to the problem of delayed maturation and the switch from an ape to a human pattern of growth (see Shea 1987), this can be best achieved by looking at various growth trajectories for the fossil hominids and the energetic costs that these incur.

The principal question addressed here is whether the increasing energetic costs associated with the expansion of the brain may have pushed hominid growth patterns in the direction of delayed maturation and more altricial neonates, and if so, when is this likely to have occurred? An initial assumption consistent with basic evolutionary principles would be that it is advantageous to the mother to minimize birth interval to maximize lifetime reproductive success. If this holds, then mothers would benefit from rapidly growing young. The fact that hominids have evolved in the opposite direction – towards longer birth intervals and longer periods of infant dependence – suggests that females have been constrained from higher reproductive rates. The hypothesis here is that the energetic requirements of the brain are imposing this constraint. The problem, then, is to quantify these constraints to assess where in hominid evolution secondary altriciality may have evolved.

The simple assumption employed in the analyses set out below is that the absolute brain costs of the human growth curve represent a ceiling for hominids. Table 2 shows the energy requirements for brain maintenance of various hominid taxa during the first five years of life. The energy costs are calculated from the adult brain mass assuming various growth rates. The first set of assumptions is that growth is similar to that of chimpanzees (i.e. 47 % neonate brain mass, 75.7 % at year 1, and so on). As can be seen, the energy requirements for all species lie below those of the modern human, and for the neonate, even the largest brained australopithecine – *A. boisei* – has costs that are only 65 % of that of modern humans. Although these costs are greater than those of chimpanzees, it may be argued that there is little by way of selective pressures for the evolution of altriciality in this genus. It may perhaps be inferred that at this stage of hominid evolution there had been no significant shift in growth patterns; an inference broadly consistent with other lines of evidence (Bromage & Dean 1985; Bromage 1987). For the genus *Homo*, on the other hand, costs begin to exceed those of modern humans. The neonate requirements of all species are greater than those of modern humans with their slower rates of growth, and for later hominids this effect spreads through to the older age groups. It thus may be argued that with the larger brains associated with the genus *Homo* – that is above 600 cm^3 – some shift towards altriciality should occur if energy is a significant constraint.

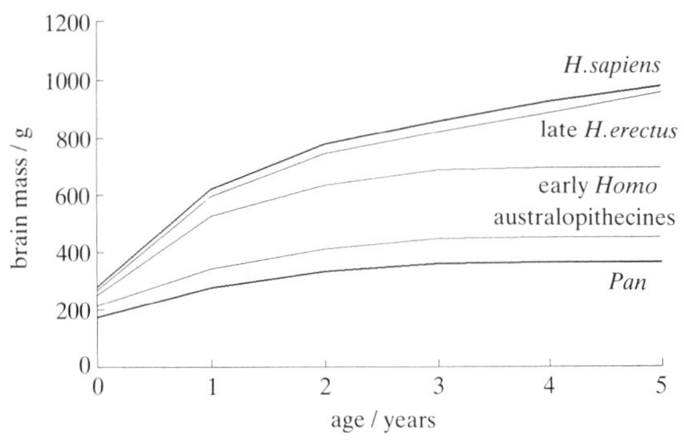

Figure 3. Predicted pattern of brain growth for fossil hominids compared with chimpanzees and modern humans. See text for discussion of methods.

This approach can be extended to model the effects on energy requirements of changing growth rates. For example, Table 2*b* shows the energy requirements for brain maintenance if growth rates are shifted 25 % of the way from chimpanzee to modern human patterns. As can be seen, as the growth rates become more retarded the brain costs of various hominid taxa drop below that of modern humans. Table 2*c*, *d* shows the effects of further shifts – 50 % and 75 % – towards modern humans.

From these simulations it is possible to construct some expectations for the growth patterns of specific hominid taxa. First, it can be argued that none of the non-*Homo* species would have departed significantly from the ancestral ape pattern in their growth rates. Second, the shift from an ape to a human growth pattern would have occurred initially in the earliest phase of infancy, with only a gradual effect on later phases. In other words, hominid growth shifted first among neonates and then successively through the later phases of infancy. Third, the appearance of genus *Homo* is likely to coincide with the first shift in growth rates towards a human pattern. And fourth, the fully modern human growth patterns are likely to have become established during the course of *Homo erectus*. The predicted growth curves of the hominid taxa are shown in figure 3, based on an assumption of minimum change from chimpanzees compatible with an energetic cost below that sustained by modern humans.

Putting these models into a more specific historical context, it can be argued that the critical threshold for a change in neonate brain size from the ape baseline is 590 g adult brain mass; for a change in first year growth rate the threshold is 820 g adult brain mass; and for years 2, 3, 4 and 5 it is 860, 870, 980 and 980 g respectively. In evolutionary terms these values coincide with *H. habilis* for the neonate, and various temporal successive stages of *H. erectus* (*sensu lato*) for the later ages. By the time of *H. sapiens sensu lato* it would be expected that fully modern growth rates would occur. This is consistent with growth patterns inferred from dental development in neanderthals (Stringer *et al.* 1990; Trinkaus & Tompkins 1990).

6. HOMINID FORAGING AND LIFE-HISTORY STRATEGIES IN THE CONTEXT OF ENCEPHALIZATION

The conclusion to be drawn from the analyses and data presented above is that there is an ecological dimension to encephalization. The precise implications for hominid foraging behaviour should therefore be considered.

(*a*) *Timing of dietary shifts*

A number of foraging and dietary changes have been proposed for hominids during the course of their evolution: meat eating, increased extractive efficiency through technology, use of underground resources, food sharing, central place foraging. If the higher costs of encephalization are the primary causes of these shifts then there is no reason to expect these to occur before the appearance and evolution of the genus *Homo*. Andrews & Martin (this symposium) have shown that there is very little evidence for major changes in diet among the hominoids before 2.0 Ma. The appearance of stone tools, evidence for large mammal bone processing (Blumenschine, this symposium), and the change in hominid dentition around this date, coincident with the origins of genus *Homo*, may indicate that encephalization is closely related to changes in ecology.

(*b*) *Additional energetic requirements*

The central point here is that having a large brain imposes additional energetic costs on both the infant and the mother. As discussed above, brain costs are only a proportion of overall energy requirements, but using the same methodology we can suggest the extent to which various hominid taxa will have required greater energy inputs for offspring. Figure 4 shows the percentage of additional energy requirements above those of a chimpanzee, by using modern human infant energetics data (Ulijaszek & Strickland 1991) less the brain costs, plus the estimated brain costs for a series of modelled levels of encephalization. Again it can be seen that australopithecines (basal hominids) have costs only about 1% higher than chimpanzees, but that for *Homo* the additional costs vary from 5 to 17%, depending upon brain size and age. It could be argued that these models indicate the extent to which different hominids require different levels of energy intake.

(*c*) *Meat eating*

These estimates suggest that for all *Homo* species there is a significant additional energetic requirement, but perhaps the most interesting point is that they are not higher. The recent debate over the importance of meat-eating in human evolution has focussed closely on the problem of the means of acquirement (see Blumenschine, this symposium), but rather less on the quantities involved. The implications of the data presented here is that a small increase in the level of energy and protein available through meat eating may have major consequences for the nutritional status of the populations concerned. Given that other plant foods, especially underground resources, may also be incorporated into the diet, then a small increase in the amount of scavenging and hunting over the levels observed in chimpanzees may be sufficient. In considering the evolution of human carnivory it may be that a level of 10–20% of nutritional intake may be sufficient to have major evolutionary consequences, and that the conditions under which protein starvation (Speth, this symposium) may result will seldom occur.

(*d*) *Costs and benefits of resource exploitation*

Hawkes (this symposium) has argued that selection of resources is not merely a function of availability and nutritional quality, but of these in relation to the costs of exploitation and processing. Meat eating, it may be argued, represents an expansion of resource breadth beyond that found in non-human primates. What may be equally significant is that hominids, through technology or some other aspect of their behaviour such as cooperative foraging, may have increased returns in foraging by reducing costs. This may also be the case with bipedalism (Foley 1991). In other words, some early hominids had greater amounts of energy available to them because they were able to exploit resources more efficiently and at lower energetic costs. This in turn provided the energetic basis for high rates of encephalization.

(*e*) *Palaoecology*

Hominids at the time of the appearance of *Homo* were occupying the savanna environments that had expanded in the late Tertiary (Foley 1987*a*). Although these to some extent are resource depleted compared to tropical forests for a frugivore, and are highly seasonal, they do offer a different resource structure in terms of patch distribution and the abundance of large mammals that may have provided the preconditions for a change in foraging behaviour that enhanced the availability of high quality foods at critical times of the year. *Homo*, with its associated encephalization, may have been the product of the selection for individuals capable of exploiting these energy- and protein-rich resources as the habitats themselves expanded (Foley 1987*a*).

(*f*) *Maternal strategies*

Mothers from a variety of primate (and other mammalian species) have a goal of weaning infants at an optimal mass, ensuring those infants' survival, and themselves producing again (Lee *et al.* 1991). What is variable between the species are the ecological constraints on what the rate of growth to weaning can be.

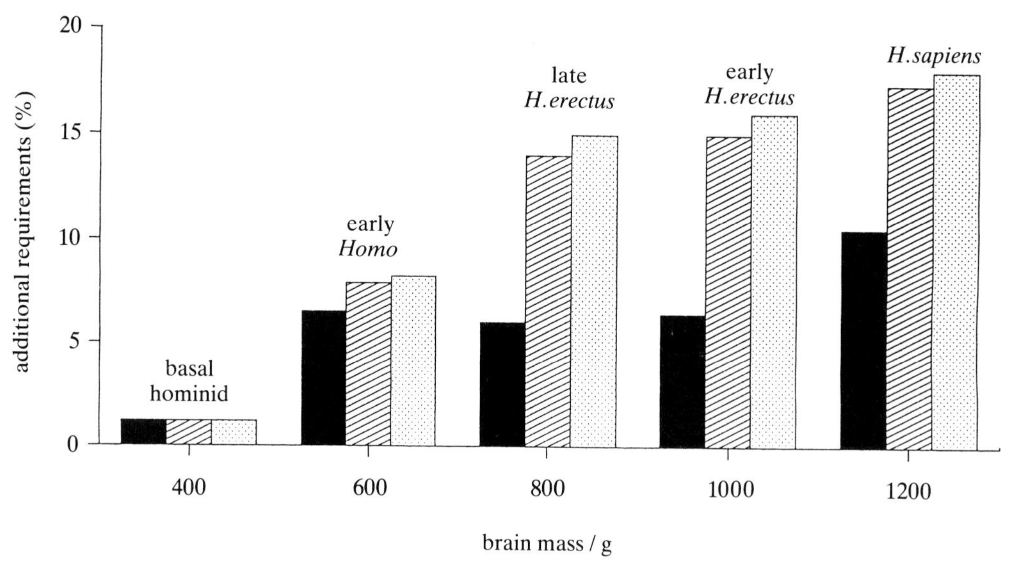

Figure 4. Estimates of the percentage increase in energy requirements of different brain sizes above those of chimpanzees for hominid infants from birth to 18 months. See text for method. Solid bars, neonate; dashed bars, 12 months; stippled bars, 18 months.

The energy available to the mother for offspring growth can either derive from the incorporation of higher quality food, from feeding for longer each day, or from maintaining lactation over a longer period. The strategy pursued by the mother results from the constraints on energy available through the diet, the time and energy costs of processing that diet, and the time budget limitations on the mother.

Among the hominoids, the relatively higher costs of growth imposed by the larger brain is met partially by the incorporation of higher energy foods (e.g. termites among female chimpanzees (McGrew 1979)). If food energy is not sufficient to maintain a 'typical' growth rate, then lactation is extended, with consequences for correlated life-history parameters (Lee *et al.* 1991). Even with the incorporation of high quality energy from animal foods, the opportunity and foraging costs associated with animal foods might have limited the mother's ability to sustain high growth rates, leading to selection for an energetically less demanding strategy of increase the duration of growth.

The observation of slower rates of development and an extended period of immaturity among *Homo* suggests that mothers were indeed under energetic and time-budget constraints which could not be compensated for entirely by a shift to a higher quality diet. It also suggests that alternatives such as male provisioning or utilizing female kin caretakers to lessen the mother's energetic burden were unlikely to play a major role in maintaining high growth rates and rapid development, at least in the earlier *Homo*. As a result there was selection for a shift in life-history parameters towards slower development.

7. CONCLUSIONS

Whereas bipedalism has generally been considered in ecological terms, encephalization in hominid evolution has not received the same treatment (Milton 1988). The link between relative brain size and metabolic rate (Martin 1981, 1983; Parker 1990) has provided the theoretical basis for examining how the evolution of larger brain size relates to changes in foraging and dietary behaviour, as well as to changes in life-history parameters. It has been argued here that the increased energy requirements associated with larger-brained offspring imposed additional costs. Simulation of these costs indicates that mothers would have been required to raise the level of energetic maternal effort by between 1 and 17% during the course of hominid evolution. One means of achieving this would be through slower growth rates, which in turn would have constrained the lifetime reproductive success of female hominids. This change in life-history strategy appears to be associated specifically with the genus *Homo*. It is probable that the evolution of larger brains within the genus requires both increased foraging efficiency and the incorporation of higher quality foods in substantial quantities in the diet, in combination with the life-history changes. The conclusion to be drawn is that encephalization among hominids, or indeed among other mammalian lineages, is dependent upon ecological conditions.

REFERENCES

Aiello, L. C. & Dean, M. C. 1990 *An introduction to human evolutionary anatomy*. London: Academic Press.

Beynon, A. D. & Dean, M. C. 1988 Distinct dental development patterns in early fossil hominids. *Nature, Lond.* **335**, 509–514.

Beynon, A. D. & Wood, B. A. 1988 Patterns and rates of molar crown formation times in East African hominids. *Nature, Lond.* **326**, 493–496.

Binford, L. R. 1984 *Faunal remains from Klasies river mouth.* New York: Academic Press.

Bromage, T. G. 1987 The biological and chronological maturation of early hominids. *J. hum. Evol.* **16**, 257–272.

Bromage, T. G. & Dean, M. C. 1985 Re-evaluation of the age at death of Plio-Pleistocene fossil hominids. *Nature, Lond.* **317**, 525–528.

Byrne, R. & Whiten, A. eds. 1988 *Machiavellian intelligence.* Oxford: Clarendon.

Clutton-Brock, T. H. & Harvey, P. H. 1980 Primates, brains and ecology. *J. Zool., Lond.* **190**, 309–323.

Davies, A. G., Caldecott, J. O. & Chivers, D. J. 1984 Natural foods as a guide to nutrition of Old World primates. In *Standards in laboratory animal management*, pp. 225–244. Potters Bar, Hertfordshire: Universities Federation for Animal Welfare.

Foley, R. A. 1987*a* *Another unique species: patterns in human evolutionary ecology*. Harlow: Longman.

Foley, R. A. 1987*b* Hominid species and stone tool assemblages: how are they related? *Antiquity* **61**, 380–392.

Foley, R. A. 1990 The causes of brain enlargement in human evolution. *Behav. Brain Sci.* **13**, 354–356.

Foley, R. A. 1991 The evolutionary ecology of fossil hominids. In *Ecology, evolution and human behavior* (ed. E. Smith & B. Winterhalder). University of Chicago Press. (In the press.)

Gibson, K. R. 1986 Cognition, brain size and the extraction of embedded food resources. In *Primate ontogeny, cognition and social behaviour* (ed. J. G. Else & P. C. Lee). pp. 93–105. Cambridge University Press.

Harvey, P. H., Martin, R. D. & Clutton-Brock, T. H. 1987 Life histories in comparative perspective. In *Primate societies* (ed. B. B. Smuts, D. L. Cheney, R. M. Seyfarth, R. W. Wrangham & T. T. Struhsaker). pp. 181–196. University of Chicago Press.

Hladik, C. M. 1978 Adaptative strategies of primates in relation to leaf-eating. In *The ecology of arboreal folivores*. (ed. G. G. Montgomery), pp. 373–395. Washington D.C.: Smithsonian Institution Press.

Humphrey, N. K. 1976 The social function of intellect. In: *Growing points in ethology* (ed. P. P. G. Bateson & R. A. Hinde), pp. 303–317. Cambridge University Press.

Jerison, H. J. 1973 *Evolution of the brain and intelligence*. New York: Academic Press.

Jolly, A. 1966 Lemur social behaviour and primate intelligence. *Science, Wash.* **153**, 501–506.

Kleiber, M. 1961 *The fire of life*. New York: John Wiley.

Lee, P. C. 1989 Comparative ethological approaches in modelling hominid behaviour. *Ossa* **14**, 113–126.

Lee, P. C., Majluf, O. & Gordon, I. J. 1991 Growth, weaning and maternal investment from a comparative perspective. *J. Zool., Lond.* **225**, 99–114.

Martin, R. D. 1981 Relative brain size and basal metabolic rate in terrestrial vertebrates. *Nature, Lond.* **293**, 57–60.

Martin, R. D. 1983 *Human brain evolution in an ecological context*. 52nd James Arthur Lecture on the Evolution of the Brain. American Museum of Natural History.

Martin, R. D. 1989 *Primate origins and evolution: a phylogenetic reconstruction*. London: Chapman and Hall.

McGrew, W. C. 1979 Evolutionary implications of sex differences in chimpanzee predation and tool use. In *The great apes* (ed. D. A. Hamburg & E. R. McCown) pp. 440–463. Menlo Park: Benjamin Cummings.

McHenry, H. M. 1988 New estimates of body weight in early hominids and their significance to encephalization and megadontia in 'robust' australopithecines. In *Evolutionary history of the 'robust' australopithecines* (ed. F. E. Grine) pp. 133–148. Chicago: Aldine de Gruyter.

McNab, B. K. 1978 Energetics of arboreal folivores: physiological problems and ecological consequences of feeding on an ubiquitous food supply. In *The ecology of arboreal folivores* (ed. G. G. Montgomery) pp. 153–162. Washington, D.C.: Smithsonian Institution Press.

Milton, K. 1984 The role of food processing factors in primate food choice. In *Adaptations for foraging in non-human primates* (ed. P. S. Rodman & J. G. H. Cant) pp. 249–279. New York: Columbia University Press.

Milton, K. 1988 Foraging behaviour and the evolution of primate intelligence. In *Machiavellian intelligence* (ed. R. W. Byrne & A. Whiten) pp. 285–305. Oxford: Clarendon Press.

Ogawa, T., Kamiya, T., Sakai, S. & Hosokawa, H. 1970 Some observations on the endocranial cast of the Amud man. In *The Amud man and his cave site* (ed. H. Suzuki & F. Takai) Tokyo: Academic Press of Japan.

Parker, S. T. 1990 Why big brains are so rare? In '*Language' and intelligence in monkeys and apes: comparative developmental perspectives* (ed. S. T. Parker & K. R. Gibson) pp. 129–154. Cambridge University Press.

Parker, S. T. & Gibson, K. R. 1979 A model of the evolution of intelligence and language in early hominids. *Behav. Brain Sci.* **2**, 367–407.

Passingham, R. 1982 *The human primate*. San Fransisco: Freeman.

Passmore, R. & Durnin, J. B. G. 1955 Human energy expenditure. *Physiol. Rev.* **35**, 801–835.

Rasmussen, D. T. & Izard, M. K. 1988 Scaling of growth and life history traits relative to brain size and metabolic rate in lorises and galagos. *Am. J. phys. Anthrop.* **75**, 357–367.

Richard, A. 1985 *Primates in nature*. San Francisco: Freeman.

Rodman, P. S. & McHenry, H. M. 1980 Bioenergetics and origins of bipedalism. *Am. J. phys. Anthrop.* **52**, 103–106.

Schultz, A. H. 1969 *The life of primates*. London: Weidenfield and Nicholson.

Shea, B. T. 1987 Reproductive strategies, body size and encephalization in primate evolution. *Int. J. Primatol.* **8**, 139–156.

Smuts, B. B., Cheney, D. L., Seyfarth, R. M., Wrangham, R. W. & Struhsaker, T. T. (eds) 1987 *Primate societies*. University of Chicago Press.

Standen, V. & Foley, R. A. (eds) 1989 *Comparative socio-ecology: the behaviouralecology of humans and other mammals* (Special Publications of the British Ecological Society **8**). Oxford: Blackwell Scientific Publications.

Stringer, C. B., Dean, M. C. & Martin, R. D. 1990 A comparative study of cranial and dental development within a recent British sample and among neandertals. In *Primate life history and evolution* (ed. C. J. DeRousseau) pp. 115–152. New York: Wiley-Liss.

Torrence, R. 1983 Time-budgeting and hunter-gatherer technology. In *Hunter-gatherer economy in prehistory: a European perspective* (ed. G. N. Bailey) pp. 11–22. Cambridge University Press.

Trinkaus, E. & Tompkins, R. L. 1990 The neandertal life-cycle: the possibility, probability and perceptibility of contrasts with recent humans. In *Primate life history and evolution* (ed. C. J. DeRousseau) pp. 153–180. New York: Wiley-Liss.

Ulijaszek, S. J. & Strickland, S. 1991 *Nutritional Anthropology*. Cambridge University Press.

Walker, A. C., Leakey, R. E., Harris, J. M. & Brown, F. H. 1986 2.5 Myr *Australopithecus boisei* from west of Lake Turkana. *Nature, Lond.* **322**, 517–522.

Waser, P. M. & Homewood, K. 1979 Cost-benefit approaches to territoriality: a test with forest primates. *Behav. Ecol. Sociobiol.* **6**, 115–119.

Waterman, P. G. 1984 Food acquisition and processing as a function of plant chemistry. In *Food acquisition and processing in primates* (ed. D. J. Chivers, B. A. Wood & A. Bilsborough), pp. 177–211. New York: Plenum Press.

Wrangham, R. W. 1986 Ecology and social relationships of two species of chimpanzee. In *Ecological aspects of social evolution* (ed. D. I. Rubenstein & R. W. Wrangham), pp. 352–378. Princeton University Press.

Discussion

E. M. Widdowson (*9 Boot Lane, Barrington, Cambridge, U.K.*). Dr Foley has discussed only the brain as a whole. If he is interested in intelligence it seems as though it should be the forebrain he should be considering. Has he any information about the forebrain of gorillas and chimpanzees compared with that of man?

R. A. Foley. It is certainly the case that certain parts of the forebrain in humans have expanded at a greater rate than the other parts. This is especially true of the neocortex, although it should be noted that the cerebellum has also increased at a higher rate. The neocortex is 3.2 times larger than predicted from non-human primates based on body size alone (Passingham 1982). However, it appears that this allometric pattern of brain part increase is general to the primates and mammals as a whole. In other words, a plot of neocortex volume against brain volume yields a very strong log-linear relationship, showing that species with large brains have larger neocortex. Modern humans fall on the predicted line for size of neocortex for their overall brain volume. This suggests that total brain size is a useful indicator of other neocortex size.

E. M. Widdowson. Does a change to bipedalism involve an increase in the calcium requirement for the longer and wider bones?

R. A. Foley. It is difficult to answer the question about calcium requirements on a *pro rata* basis, but assuming that chimpanzee and gorilla bones require the same levels of calcium as humans, we can ask whether humans have more skeletal mass per unit body mass.

There is an allometric relationship between body mass and skeletal mass (Prange *et al.* 1979), although this has not been calculated in detail for primates. However, the principal parts of the skeleton represent 79% of body mass in chimpanzees, 65% in humans and 51% in gorillas (Steudel 1980). Given this scaling with body mass, there is little reason to expect higher requirements associated with bipedalism.

References

Prange, H. D., Anderson, J. F. & Rahn, H. 1979 Scaling of skeletal mass to body mass in birds and mammals. *Am. Nat.* **113**, 103–122.

Steudel, K. 1980 New estimates of early hominid body size. *Am. J. phys. Anthrop.* **52**, 63–70.

C. D. Knight (*Polytechnic of East London, U.K.*) Dr Foley stressed the high energetic costs of large brain size, and the extent to which these costs would have posed problems for mothers having to feed and care for large-brained offspring. He then went on to picture bipedalism as the factor that enabled evolving females to cope with these burdens. How closely is he linking encephalization with what seems to be the much earlier emergence of bipedalism? Dr Foley will be aware that there are other models that make no such connection, linking encephalization instead with the success of females in increasingly harnessing the provisioning energies of males; features such as ovulation-concealment and continuous receptivity being interpreted as functional in this context. How far would he go in arguing that the bipedal female solved her own problems without need for a change in mating strategies?

R. A. Foley. The improvement in the fossil record in recent years has shown us that, chronologically, bipedalism and encephalization are disjoint. It may well be that bipedalism is a necessary (but not sufficient) precondition for major encephalization. The simple point can be made that bipedalism is in some circumstances an energetically efficient adaptation, and therefore this might counterbalance some of the additional requirements of encephalization (Foley 1991). Dr Knight is right to point out that female strategies may evolve independently. What we are less certain about is whether major changes in the parenting behaviour could occur without some shift in the mating strategy, as by and large these two aspects are closely linked.

J. H. P. Jonxis (*Rijksstraatweg 65, 9752 AC Haren, The Netherlands*). In the first months after birth the energy metabolism of the brain is far higher than in a somewhat older baby. In those first months the brain covers about half of the total heat production.

R. A. Foley. This is an illuminating comment and underpins the argument we have put forward that brains impose high costs which will be increased if encephalization occurs.

Traditional diet and food preferences of Australian Aboriginal hunter–gatherers

KERIN O'DEA

Department of Human Nutrition, Deakin University, Geelong, Victoria 3217, Australia

SUMMARY

Until European settlement of Australia 200 years ago, Aborigines lived as nomadic hunter–gatherers all over the continent under widely varying geographic and climatic conditions. Successful survival depended on a comprehensive knowledge of the flora and fauna of their territory. Available data suggest that they were physically fit and lean, and consumed a varied diet in which animal foods were a major component. Despite this, the diet was not high in fat, as wild animal carcasses have very low fat contents through most of the year, and the meat is extremely lean. Everything on an animal carcass was eaten, including the small fat depots and organ meats (which were highly prized), bone marrow, some stomach contents, peritoneal fluid and blood. A wide variety of uncultivated plant foods was eaten in the traditional diet: roots, starchy tubers, seeds, fruits and nuts. The plant foods were generally high in fibre and contained carbohydrates, which was slowly digested and absorbed. Traditional methods of food preparation (usually baked whole or eaten raw) ensured maximum retention of nutrients. In general, traditional foods had a low energy density but high density of some nutrients. The low energy density of the diet and the labour intensity of food procurement provided a natural constraint on energy intake. This, together with the other nutritional qualities of the diet (including high fibre, slowly digested carbohydrate, very low saturated fat, relatively high proportion of the long-chain highly polyunsaturated fatty acids, low sodium and high potassium, magnesium and calcium) would have protected against obesity, non-insulin-dependent diabetes, and cardiovascular diseases, all of which are highly prevalent in westernized Aboriginal communities in Australia today.

1. INTRODUCTION

Prehistorians believe that Aborigines came to Australia from South East Asia at least 40000–50000 years ago (Flood 1983). They were probably then a coastal people, and are believed to have settled first on the northern coast of Australia, before moving slowly south to other parts of the continent. In more recent times, and until European colonization of Australia just over 200 years ago, Aborigines lived as hunter–gatherers all over the continent under widely varying geographic and climatic conditions, ranging from the tropical coastal regions of the north (latitude 11°–20° S), through the vast arid regions of the Centre (latitude 20°–30° S), to the cool–temperate regions of the south (latitude 30°–43° S). The more fertile coastal areas, both north and south, could sustain larger populations than the arid inland or desert areas. Little reliable information is available about the Aborigines of southeastern Australia as their land was the first to be taken over by European settlers for farming, with the Aboriginal population being decimated by battle and introduced diseases such as smallpox. However, it is clear from recent studies of traditional vegetable foods (Brand *et al.* 1983) and the continuing abundance of wild animals (Naughton *et al.* 1986) in uncleared land, that this part of Australia could have supported the greatest population density of Aborigines, consistent with the current concentration of the European-derived population along the southeastern seaboard. The arid inland regions supported much smaller numbers of people. Each tribal group hunted and gathered food in a defined territory that could be as vast as 100000 km^2 in the desert regions or as small as 500 km^2 in fertile coastal country (Kirk 1981).

2. ABORIGINAL HEALTH PRE-WESTERNIZATION

Although there is little reliable, quantitative data on diseases in Aborigines with little or no European contact (Basedow 1932), numerous early reports described them as having been lean and apparently physically fit (see table 1). Undoubtedly, the nomadic lifestyle and its associated customs promoted survival of the fittest. In the most remote areas of Australia, small groups of Aborigines continued to live a nomadic lifestyle until 20–30 years ago, and data collected from such groups suggest they were very lean and had no evidence of the chronic diseases that occur in epidemic proportions in westernized Aboriginal communities today (Elphinstone 1971). For example, a group of adult Aborigines from the Great Sandy Desert in

Table 1. *Health of Australian Aborigines as hunter–gatherers*

physically fit
lean (body mass index (BMI) < 20 kg m^{-2})
low blood pressure
no age-related increase in BMI or blood pressure
low fasting glucose
low fasting cholesterol
no evidence of diabetes or coronary heart disease

Western Australia who were examined at 'first contact' in the 1960s by a medical officer had body mass indexes (BMI) ranging from 13.2 to 19.5 kg m^{-2} (Elphinstone 1966). Abbie (1971) reported low BMI (mean of 19–20 kg m^{-2} for the men, and 18–20 kg m^{-2} for the women) and no increase with age in full-blood Aborigines 'living under nomadic or near nomadic conditions' in northern Australia between 1959 and 1963. Indeed, the women showed a trend to reduced body mass with increasing age, consistent with the observations of others that Aboriginal women living traditionally showed 'an obvious loss of subcutaneous tissue beyond early adult life' (Abbie 1966). More recently, as part of a long-term study on the relation between lifestyle and health in small traditionally oriented Aboriginal communities in northeast Arnhemland, White (1985) also reported a relative loss of body mass and subcutaneous fat in the older women compared with the older men. Both Abbie (1966, 1971) and White (1985) have reported low resting blood pressures in adult Aborigines living a traditionally oriented lifestyle and no increase in blood pressure with age in the 20–50 year age group. These observations are consistent with the findings of Lowenstein (1961) in other populations living traditionally, and support his conclusion that increasing blood pressure is not an inevitable consequence of aging, but rather a consequence of western lifestyle.

As part of a long-term study by White (1985) and colleagues on the relation between lifestyle and health in small traditionally oriented Aboriginal groups in northeast Arnhemland, we recently examined several biochemical parameters in the fasting state in one of these outstations, the details of which have been reported previously (O'Dea *et al.* 1988*b*). By standard criteria for BMI, all of the adults were 'underweight', with BMI ranging from 13.4–19.3 kg m^{-2}. Despite this, they displayed no biochemical evidence of malnutrition; none were anaemic and red cell folate levels were all within the normal range for European Australians (a unique observation in itself). In addition, their fasting glucose (3.8 ± 0.4 mmol l^{-1}) and cholesterol concentrations (3.9 ± 0.2 mmol l^{-1}) were low relative to urbanized Aborigines and Caucasians. However, their fasting insulin levels (13 ± 4 mU l^{-1}) were similar to other more westernized young Aboriginal men and higher than those of Caucasian men, both having higher BMIs (approximately 21 kg m^{-2}) (O'Dea *et al.* 1982). These data suggest that, in view of their low BMI (approximately 17 kg m^{-2}) and low fasting glucose levels, their insulin levels were inappropriately elevated. Their fasting triglyceride concentrations (1.13 ± 0.09 mmol l^{-1}), although within the normal range for Caucasians, were higher than would be expected because of their extreme leanness, but consistent with insulin resistance. This indirect evidence for insulin resistance in these people, despite their extreme leanness, regular physical activity (daily hunting and foraging), and traditionally oriented diet suggest that, in common with other Aboriginal communities all over Australia, they may become susceptible to obesity and non-insulin-dependent diabetes mellitus (NIDDM) if they westernize further.

3. HUNTER–GATHERER LIFESTYLE

Most information on the traditional lifestyle of Australian Aborigines has come from the study of groups who continued living as hunter–gatherers well into the 20th Century, in the most remote and, to most white Australians, inhospitable parts of the country in the centre and north of the continent (White 1985; McArthur 1960; Tonkinson 1978; Hiatt & Jones 1988). These studies have produced much detailed information specific to individual groups of Aborigines, but have also revealed a number of quite striking similarities between groups which are a function of the nomadic hunter–gatherer lifestyle.

Aborigines from all over Australia were omnivorous, deriving their diet from a wide range of uncultivated plant foods and wild animals. The composition and diversity of the food supply, and the relative proportions of plant and animal foods, were greatly influenced by both the season of the year and the geographic location. Wherever Aborigines lived in Australia, successful survival depended on an intimate and detailed knowledge of the land, sources of fresh water, and the impact of the annual cycle of seasonal changes on the flora and fauna of their territory (White 1985; Hiatt & Jones 1988; Kirk 1981).

Traditionally, Aborigines lived in bands based on extended family groups usually numbering 20–30 including children. Larger tribal gatherings for traditional ceremonies were only possible if there was sufficient food available to support the larger group. Notable instances where this occurred regularly were in the alpine region of southeastern Australia during the summer months, when the bogong moth provided a plentiful food supply (Flood 1980), in southern Queensland when the nuts from the Bunya pine were ripe, and historically along the Murray river in southeastern Australia where hundreds of Aborigines feasted on freshwater crayfish or fish at different times of the year (Sahlins 1972).

Men and women both contributed importantly, but differently, to food procurement (Tonkinson 1978; Kirk 1981; White 1985). In general, women provided the subsistence diet, gathering plant foods, honey, eggs, small mammals, reptiles, fish, shellfish, crustaceans and insects. They usually hunted in groups, which allowed them to share child-minding and have an enjoyable and sociable time. Although the men participated in these activities from time to time, they were primarily hunters and provided the less regular, but highly valued 'feasts' from large mammals (such as kangaroos), birds (such as emu), reptiles (such as turtle or perente) or fish. Men usually hunted alone or in pairs, although there are reports of group hunting – for example, ambushing animals by fire.

Although it has been argued that hunter–gatherers spent less time on average each day ensuring their livelihood (3–5 h) than agriculturalists or the employed in societies such as our own (Sahlins 1972), food procurement and preparation for Aboriginal hunter–gatherers were energy-intensive processes that could involve sustained physical activity: walking long distances, digging in rocky ground for tubers deep below the surface, digging for reptiles, eggs, honey ants and witchetty grubs, chopping with a stone axe (for honey, grubs etc), winnowing and grinding of seeds, digging pits for cooking large animals, and gathering wood for fires (for cooking and warmth).

As a rule, there was one main meal in the day, in the late afternoon when people returned to camp after the day's activities. However, people would eat snacks throughout the day while hunting and gathering: grubs, fruits, gum, honey ants and sugar-bag (honey from the wild bees). Hunters would sometimes cook and eat the liver of a kangaroo before carrying the carcass back to camp. Shellfish and fish were also often cooked and eaten on the spot or, in some circumstances, eaten raw.

4. AGRICULTURE

Although Aborigines did not practise 'agriculture' in the conventional sense of the word, there is no doubt that they interacted with their environment in ways designed to ensure food abundance.

Aborigines in different parts of Australia had practices that could be described as a semi-agricultural approach to several important plant foods (Flood 1983; Kirk 1981). In northern Australia, the long yam (*Dioscorea transversa*) is a dry-season staple, and there are numerous reports of the top of the tuber being left attached to the tendril of the vine when the rest of the tuber was dug out, to ensure that the yam would grow again in the following year. Yams were also planted on off-shore islands, presumably as a 'reserve' food supply.

Fruit seeds spat out at campsites, especially if they landed on the 'compost' of other food debris, may well have been the source of native fruit trees frequently found at regularly used camps.

The seed-collecting practices of Aborigines in the Darling River basin of western New South Wales have been thoroughly documented (Allen 1974). Although they relied on the river to supply their diet for much of the year (fish, shellfish, birds, aquatic plants), in the summer the staple of their diet was the seeds of the native millet. They developed an ingenious method of harvesting, and would gather the green grass before the seed was ripe, leave it in stacks for the seeds to ripen and dry, and then thresh it and collect the seeds. Archaeological evidence shows that seed grinding in the Darling River basin goes back 15000 years (Flood 1983).

'Firestick farming' is the term used to describe the systematic burning done by Aborigines all over Australia (Jones 1980). Although it was sometimes used as a means of ambushing animals, the long-term consequences were more significant: the promotion of new growth, which both provided food directly and attracted game to the area. There is evidence that fire facilitated the germination of certain seeds. It is also believed to have maintained the vast savannah grasslands in northern Australia, and it has been suggested that fire may have been a critical factor in altering the ecosystem and extending man's habitat (Jones 1980).

It is now recognized that human intervention in the arid zone of central Australia was integral to the maintenance of the fairly fragile ecosystem around many waterholes. Aborigines kept the waterholes patent, and thereby contributed to the survival of the plant, bird and animal life, which depended upon a local water supply.

Reports from both northern and southern Australia refer to the use of a variety of strategies to trap or poison fish (Kirk 1981). For example, sophisticated systems of channels and dams were used to trap fish between the tides; and, in an imaginative hunting strategy, fish in river pools were stunned with poisons derived from plants (usually the root): people in the group could then easily harvest the fish floating unconscious on the surface, but would then avoid drinking water from the poisoned pool for several days, although they were apparently unaffected by eating the fish.

5. ANIMAL FOODS

A comprehensive knowledge of animal distribution and behaviour was essential to the success of Aborigines as hunters and gatherers (Hiatt & Jones 1988; White 1985; Tonkinson 1978; Kirk 1981). They needed to know when mammals, reptiles, birds and fish were most likely to be 'fat', their hibernation and nesting behaviour, where and when eggs were laid, the most propitious time (in the day or year) to find a particular species and where to find it. For example, they knew that the platypus laid eggs, but were not believed by English naturalists! Another illustration of the depth and detail of their knowledge of animals was the complex system of names they developed for species that were important food sources, frequently differentiating between genders (e.g. kangaroo) or different phases of the lifecycle (e.g. barramundi, a prized fish in northern Australian rivers and coastal waters). In contrast, many small birds that were not part of the normal diet were grouped together under a common name that translates as 'small bird' (Hiatt & Jones 1988).

All animals were potential food sources: mammals, birds, reptiles, insects and marine species (mammals, reptiles, fish, crustacea and shellfish). Everything edible on an animal carcass was eaten, including muscle, fat depots and internal organs (although usually not intestinal contents). One of the most striking characteristics of wild (or non-domesticated) animals is the low fat content of their carcasses (Naughton *et al.*

1986). In this they differ strikingly from domesticated meat animals such as cattle, sheep, pigs and chickens which have fat deposits under the skin, within the abdomen and between and within muscles. Wild animals do have discrete depots of fat within the abdomen (primarily surrounding the gonads, kidneys and intestines) and these depots frequently increase in size at particular times of the year. However, they generally tend to be small and usually have to be shared among many people. The meat (muscle) is always lean irrespective of the season and does not 'marble'. Because of this, a high proportion of what little fat is present is structural fat: part of the membranes of the muscle cells, and relatively rich in the long-chain highly polyunsaturated fatty acids (Naughton *et al.* 1986). These polyunsaturated fatty acids have important physiological functions. In addition to being vital components of all cell membranes in the human body, they are essential for normal growth and development of the brain and the retina, and are precursors of a group of chemicals (the prostaglandins) which affect numerous bodily functions including the modulation of blood flow and thrombosis (Neuringer *et al.* 1984; Sanders 1985).

Although on a mass basis, muscle constituted the major edible portion of an animal carcass, fat depots and organs (including emptied intestines and stomach) were also consumed. The organ meats (particularly brain and liver) were rich sources of cholesterol and fat. However, the fat in these organ meats was relatively rich in the long-chain highly polyunsaturated fatty acids (Naughton *et al.* 1986). Even the depot fat on some species was unusually rich in polyunsaturated fat (consistent with its 'softness' relative to beef tallow or mutton fat). Such foods were highly valued.

Animal foods (depot fat, organs) were not only relatively rich sources of polyunsaturated fat, but they also contained both major classes of polyunsaturated fatty acids (n-3 and n-6) in a ratio of about 1:3 (Naughton *et al.* 1986). The western diet contains predominantly the n-6 form (vegetable oils, margarines, etc.) and the ratio is about 1:12 (Sinclair & O'Dea 1990). The n-3 polyunsaturates are believed to be important in reducing the risk of thrombosis, in addition to their role in brain and retinal development.

Liver is an excellent source of iron and zinc, and also an unexpected source of nutrients usually associated with plant foods, such as vitamin C and folic acid, and may have therefore been a particularly important food at those times of the year when the diet was derived predominantly from animal foods. It may be significant nutritionally that the vitamin C in liver is in a particularly stable form and is not destroyed by cooking (O'Dea *et al.* 1987).

Insects also provided significant contributions to the diet seasonally, both directly and indirectly. Honey, from wild bees and the honey ant, provided important dietary carbohydrate in season. Witchetty grubs are still a seasonal delicacy in many parts of the country. Interestingly, although they are rich in fat and have a nutty–buttery taste, they have a fat composition very similar to olive oil (Naughton *et al.* 1986)! The bogong moth was so plentiful in the summer months in the mountains of southeastern Australia, that large gatherings of people could be supported while participating in ceremonies.

6. PLANT FOODS

The detailed and thorough knowledge the Aborigines had of the ecology of their environment allowed them to take full advantage of a wide range of plants as sources of food over the year: tuberous roots, seeds, fruits, nuts, gums and nectar. Dietary carbohydrate in the hunter–gatherer diet was derived from uncultivated plants, (tuberous roots, fruits, berries, seeds, nuts, beans) and honey. Cereal grains, the dietary staples of man since the development of agriculture, were not a major component of the traditional Aboriginal diet. The variety of plant foods available to hunter–gatherers in Australia was wide. Relative to many of their cultivated forms, wild plant foods are particularly rich in protein and vitamins. For example, a species of wild plum eaten by Aborigines in northern Australia (*Terminalia fernandiana*) has the highest vitamin C content of any known food (2–3 % wet mass) (Brand *et al.* 1982). A commonly eaten yam in northern Australia (*Dioscorea transversa*) is considerably more nutrient-dense than its modern equivalent, the potato, being higher in carbohydrate, protein, fibre, zinc and iron (Brand *et al.* 1983). The wild vegetable foods are also rich in dietary fibre, low in sodium, but rich in potassium, magnesium and calcium (Brand *et al.* 1983, 1985). The seeds that made significant contributions to the diet of Aborigines in certain seasons were not only rich in protein, but also contained significant quantities of linoleic acid. Although many of the vegetable foods are not rich fat sources (tuberous roots, leafy vegetables, fruits and berries), they nevertheless contain both n-6 and n-3 PUFA, and the ratio of n-6:n-3 was often much less than that seen in most seeds (Sinclair & O'Dea 1990).

The carbohydrate in many of these traditional foods has been shown to be more slowly digested and absorbed than the carbohydrate in equivalent domesticated plant foods (Thorburn *et al.* 1987*a*). The lower postprandial glucose and insulin levels elicited by the ingestion of these slowly digested wild plant foods may have been a factor in helping protect these populations from developing type 2 diabetes (Thorburn *et al.* 1987*b*) – a condition to which they are particularly vulnerable when they make the transition from a traditional to a western diet and lifestyle (Wise *et al.* 1976; Bastian 1979; O'Dea *et al.* 1988; O'Dea *et al.* 1990) – although other factors, in particular their low-fat, high-fibre diet, and their leanness and physical fitness, were undoubtedly also important.

Relative to their modern domesticated equivalents, the vegetable foods in traditional hunter–gatherer diets were higher in protein, fibre and vitamins, contributed PUFA from both the n-6 and n-3 series and contained carbohydrate which is slowly digested and absorbed. In general, the wild vegetable foods were bulky, with high nutrient density but low energy density (table 2). The only carbohydrate source with high energy density was wild honey.

Table 2. *Comparison of lifestyle and diet of Australian Aborigines as hunter–gatherers and after westernization*

	hunter–gatherer lifestyle	western lifestyle
physical activity level	high	low
principal characteristics of diet		
energy density	low	high
energy intake	usually adequate	excessive
nutrient density	high	low
nutrient composition of diet		
protein	high	low–moderate
animal	high	moderate
vegetable	low – moderate	low
carbohydrate	moderate (slowly digested)	high (rapidly digested)
complex carbohydrate	moderate	moderate
simple carbohydrate	usually low (honey)	high (sucrose)
fibre	high	low
fat	low	high
vegetable	low	low
animal	low (polyunsaturated)	high (saturated)
Na:K ratio	low	high

7. FOOD PREPARATION

Methods of food preparation utilized by Aborigines generally resulted in minimal loss of nutrients, particularly micronutrients. It has been estimated that about half of the plant foods eaten by Aborigines in northern Australia were processed in some way before being eaten (Beck 1985), although this is likely to have varied regionally. Many plant foods were eaten fresh and raw – frequently as they were collected – such as fruits, bulbs, nectar, gums, flowers etc. Foods were not processed unnecessarily. Any processing was done to render a food edible, more digestible or more palatable: for example, cooking of starchy tubers or seeds, grinding and roasting seeds, cooking of meat. Some food processing techniques were directed at detoxification of a potentially poisonous food (Kirk 1981; Flood 1983; Hiatt & Jones 1988; Beck *et al.* 1988). For example, the nuts of the cycad palm (*Cycas angulata*) in northern Australia were pounded into a soft pulp and then left in running water for several days, a treatment which removed the characteristic bitter taste, and also substantially removed cycasin, a known carcinogen. It is unlikely that the people concerned were aware of the potential long-term toxicity (although they counselled against eating 'too much' of the bread from the cycad nuts even when it had been thoroughly processed (Beck *et al.* 1988)) – they were primarily concerned about removing the 'bitterness'. A similar processing technique was applied to the 'cheeky' or bitter-tasting yams in northern Australia (e.g. *Dioscorea bulbifera*): they were cut up and left in running water for several days after which time the bitterness was no longer evident and they could be baked and eaten. Other bitter tubers were baked in earth ovens to break down the toxic compounds and make them taste 'sweet' (Kirk 1981; Flood 1983; Hiatt & Jones 1988; Beck *et al.* 1988).

The most common form of food preparation was cooking: roasting on the coals or baking in an earth oven. In many cases this was the only form of processing. Vegetable foods that were not eaten raw were roasted or baked whole, which protected labile micronutrients against oxidation and prevented the leaching out of water-soluble nutrients as happens to a greater or lesser degree when vegetables are chopped up, boiled and the water discarded (a common method of vegetable preparation in contemporary Australian cooking). Small animals (reptiles, birds, fish, Crustacea and other shellfish) were often cooked in a similar manner – roasted whole on the coals – with blood and other juices collecting in the peritoneal cavity and being eaten as a soup along with the organs before the meat was distributed. Such foods were usually only lightly cooked.

Larger animals (kangaroo, emu, turtle etc.) were prepared in a more complex manner, with the hunter taking responsibility (Tonkinson 1978; McArthur 1960; White 1985). From the nutritional point of view, however, the end result was similar in terms of retention of nutrients. In the case of kangaroo preparation in the Kimberley in the present day, the hunter makes a small incision in the stomach, discards intestinal contents but retains the emptied intestine and stomach, and closes the incision with a stick. Back at the camp he prepares a big fire and digs a pit. The kangaroo tail is cut off, the sinews removed and the tail cooked separately on the coals. The liver may also be roasted separately on the coals. The kangaroo has its legs broken at the shoulders and hips to facilitate thorough baking, and the feet are removed. It is then thrown on the fire and singed, which has the function of 'sealing' the cooking vessel. When the fire dies down, the animal is placed on its back in the pit and covered with hot ashes and sand. When it is judged to be cooked, the kangaroo is lifted out of the pit and placed on its back for butchering. The abdomen is opened and all present share in drinking the 'soup' in the peritoneal cavity, a mixture of blood and juices from meat and internal organs which tastes like a rich beef broth. Some organs are also eaten at this time and then the carcass is carved up, with each cut being given to particular kin of the hunter if they are in camp. Large game, like kangaroo, is always shared. The hunter keeps the head (with the brain), liver and fat depots: not much meat as such, but the richest sources of fat and polyunsaturated fat and cholesterol.

Cooking procedures for other large game are similarly ritualized. Turtle, for example, is not cooked in a pit, but placed directly on the coals. The cooking process is facilitated by placing hot stones from the fire inside the abdomen after the animal has been gutted. Not all intestinal contents are necessarily discarded from the large turtle, only those below a particular anatomical point. The intestines themselves are roasted directly on the coals and are a delicacy.

8. EATING PATTERNS

Food intake could vary enormously both on a day-to-day and a seasonal basis (O'Dea *et al.* 1988*b*; White 1985; Kirk 1981; Flood 1983). This has been described as a 'feast-and-famine' pattern of food intake. The actual pattern could probably be described more accurately as subsistence interspersed with feasts. As already noted, the women were primarily responsible for the subsistence component, and although the men did assist with these activities, they were primarily hunters, and provided the less regular, but highly valued, 'feasts'.

Food was usually consumed at the time it was available, and wastage was rare. However, in some circumstances foods were processed into forms in which they could be stored. Drying was the most usual form of processing used, and the food was either stored as such or processed further (grinding and cooking of seeds or nuts) and stored as cakes, sometimes wrapped in leaves and buried in the ground or placed in a dry and inaccessable place.

There are numerous reports of Aborigines eating 2–3 kg of meat at one long sitting, taking maximum advantage of an abundant food supply on those irregular occasions when it was available (O'Dea *et al.* 1988). It can be argued that the feasts were critical to the survival of Aborigines as hunter–gatherers as they provided excess energy which could be converted into fat and deposited as adipose tissue, thereby providing an energy reserve to tide an individual over periodic food shortages (O'Dea 1991).

9. FOOD PREFERENCES

The most highly prized components of the Aboriginal hunter–gatherer diet were the relatively few energy-dense foods: depot fat, organ meats, fatty insects and honey. In general, muscle provided the bulk of the energy from a carcass. Although many animals were actively hunted at those times of the year when their fat depots were largest (with Aborigines following the 'fat cycle' (Davis 1989)), the fat depots on most animals were usually small through most of the year and needed to be shared among many people. Thus, high fat foods were either only available seasonally, or were available in small quantities. Similarly, wild honey was available only at certain times of the year, and its procurement was often associated with high energy expenditure.

Nevertheless, it is significant that in a diet that was generally characterized by its low energy density, the foods most actively sought and most highly valued were those that had a high energy density. Clearly this was an important survival strategy. From the foregoing discussion it appears that two important components of the survival strategy of Aborigines as hunter–gatherers were to maximize energy intake and minimize energy output. The reality of the lifestyle, however, resulted in a generally low energy intake (subsistence) in combination with a relatively high energy expenditure. The scenario changed dramatically with westernization.

10. IMPACT OF WESTERNIZATION ON DIET AND HEALTH OF ABORIGINES

After westernization (see table 3), hunter–gatherer food preferences and eating behaviours have been retained by Aborigines in an environment where the availability of energy-dense foods (those rich in fat and sucrose) is no longer limited, and the energy output involved in their procurement is minimal. In the western context, Aborigines have a very high intake of sucrose (primarily as sugar in tea and from carbonated beverages) and fat (from cheap fatty meats from domesticated animals, and more recently from a wide variety of processed foods) (O'Dea *et al.* 1990). If the same eating behaviour is applied to beef and lamb as had been applied to kangaroo meat (2–3 kg at one long sitting), two to four times as much energy would be consumed: 4000–12000 kcal instead of 2000–3000 kcal! This results from the much greater fat content of meats from domesticated animals relative to wild animals (Naughton *et al.* 1986). The principle is general: if food is there, it is eaten. For example, a dozen eggs will usually not be used slowly over several days, but will be cooked up for one meal, and may even be eaten by one person! When fresh fruit arrives at a remote community – a popular event that may only occur once or twice each month – it will be sold out and eaten within a day or two.

Eating behaviour and food preferences which favoured survival in the traditional hunter–gatherer lifestyle now promote the development of obesity. Obesity occurs with alarming frequency in many westernized Aboriginal communities, with 50–80% of

Table 3. *Lifestyle-related chronic diseases in Aborigines after westernization*

android obesity
hypertension
body mass and blood pressure increase with age
non-insulin dependent diabetes mellitus
coronary heart disease
elevated triglycerides, low HDL-cholesterol levels
hyperinsulinemia (insulin resistance)

those over 35 years of age being overweight or obese (O'Dea 1991). When Aborigines gain weight, the fat is deposited centrally on the trunk – an android pattern – in both men and women (O'Dea *et al.* 1990; Rutishauser & McKay 1986). In Caucasian populations, this pattern of fat distribution has been shown to be associated with increased risk of NIDDM, coronary heart disease (CHD), hypertension, hyperinsulinemia and an abnormal lipid profile (increased levels of triglycerides and reduced HDL-cholesterol levels). These conditions frequently cluster together in one individual ('syndrome X' (Reaven 1988)) and may all be linked to the presence of insulin resistance. These 'lifestyle diseases' occur with high frequency in the most westernized Aboriginal communities (table 3) (O'Dea 1991), but are still relatively infrequent in the least westernized, traditionally oriented groups (O'Dea *et al.* 1982; O'Dea *et al.* 1988). The diabetes prevalence among adult Australians of European descent is estimated to be 3.4% (Glathaar *et al.* 1985). Crude prevalence rates two to six times higher have been reported for adults in Aboriginal communities around Australia (O'Dea 1991). However, the lower life expectancy and younger age profiles of Aboriginal communities relative to the wider Australian population invalidate direct comparisons of crude prevalence rates of age-related conditions such as NIDDM and CHD. In the 20–50 year age group, the prevalence of diabetes is ten times higher in Aborigines than in Australians of European ancestry. Although fewer data are available, the prevalence of CHD and hypertension is also considerably higher in Aborigines (Wise *et al.* 1976; Bastian 1979), and hypertriglyceridemia is a striking feature of the lipid profile in all westernized Aboriginal communities in which it has been measured (O'Dea *et al.* 1980, 1982, 1988, 1990; O'Dea 1984).

The diets consumed by Aborigines in remote areas of Australia are frequently very low in fruit and vegetables, and rely very much on the staples of flour, sugar and meat. The dietary problems are compounded by poverty, isolation, paternalism and history – the legacy of the 'stockman's diet' and lifestyle. Until the 1970s it was common in remote areas of Australia for Aboriginal men (and some women) to be employed as stockmen on cattle stations. In the early days they were not paid award wages, but did receive rations for themselves and their families, comprising flour, sugar, tea, meat and tobacco. This diet was selected as much for its keeping qualities as for its simplicity, convenience and low cost. Several innovative programmes across Australia are addressing these food supply issues in community-based programs. However, it is clear that contemporary food and nutrition problems in Aboriginal communities around Australia cannot be separated from their wider social and political context.

11. THE THERAPEUTIC IMPLICATIONS OF THE HUNTER–GATHERER LIFESTYLE

Although most Aborigines in Australia today live a sedentary, westernized lifestyle, many older individuals in remote communities retain the knowledge and ability to survive as hunter–gatherers. Collaboration with such people has allowed us to document the impact on health of a 'reverse lifestyle-change', a temporary reversion of westernized Aborigines (with all the associated contemporary health problems) to traditional hunter–gatherer lifestyle. The dramatic impact of lifestyle change on health is well illustrated by the observation that when overweight diabetic Aborigines made this temporary lifestyle transition for the brief period of seven weeks there was significant weight loss, and striking improvement in all of the abnormalities of diabetes together with a reduction in the major risk factors for coronary heart disease (O'Dea 1984; O'Dea & Sinclair 1985). A comparison of the major characteristics of the diet and lifestyle of Aborigines as hunter–gatherers and following westernization is presented in table 2.

These observations have implications not only for the prevention of obesity, diabetes and cardiovascular disease for Aborigines, but can also be applied more broadly. It has been argued that the hunter–gatherer, or palaeolithic, diet and lifestyle is the one to which we, as modern humans, are genetically programmed (Eaton & Konner 1985), as the human genetic constitution has changed little in the past 40000 years since the appearance of modern man. Thus it is to this diet and lifestyle that we should turn when seeking explanations for (and solutions to) the characteristic pattern of chronic diseases that emerges in all populations when they become more 'affluent' economically (Sahlins 1972) and adopt a sedentary, westernized way of life.

REFERENCES

Abbie A. A. 1971 Studies in physical anthropology. *Aust. Aboriginal Stud.* no. 39 (Human Biology Series no. 3). Canberra: Australian Institute of Aboriginal Studies.

Abbie A. A. 1966 Physical characteristics. In *Aboriginal man in south and central Australia*, pp. 9–45. Adelaide: Government Printer.

Allen, H. R. 1974 The Bagundji of the Darling basin: cereal gatherers in an uncertain environment. *Wld Archeol.* **5**, 309–322.

Basedow, H. 1932 Diseases of the Australian Aborigines. *J. trop. Med. Hyg.* **35**, 177–278.

Bastian, P. 1979 Coronary heart disease in tribal Aborigines – the West Kimberley Survey. *Aust. N.Z. J. Med.* **9**, 284–292.

Beck, W. 1985 Technology, toxicity and subsistence: a study of Australian Aboriginal plant food processing. Ph.D. thesis, Latrobe University, Australia.

Brand, J. C., Cherikoff, V., Lee, A. & Truswell, A. S. 1982 An outstanding food source of vitamin C. *Lancet* ii, 873.

Brand, J. C., Rae, C., McDonnell, J., Lee, A., Cherikoff, V. & Truswell, A. S. 1983 The nutritional composition of Australian Aboriginal bushfoods. 1. *Fd Technol. Aust.* **35**, 293–298.

Brand, J. C., Cherikoff, V. & Truswell, A. S. 1985 The nutritional composition of Australian Aboriginal bushfoods. 3. Seeds and nuts. *Fd Technol. Aust.* **37**, 275–279.

Davis, S. 1989 *Man of all seasons*. London: Angus & Robertson.

Eaton, S. B. & Konnor, M. 1985 Paleolithic nutrition. A consideration of its nature and current implications. *New Engl. J. Med.* **312**, 283–289.

Elphinstone, J. J. 1966 Endemic diseases in natives (Kimber-

leys). In *Report of Commissioner of Public Health Western Australia, Appendix IV*. Perth: Government Printer.
Elphinstone, J. J. 1971 The health of Australian Aborigines with no previous association with Europeans. *Med. J. Aust.* **2**, 293–301.
Flood, J. 1980 *The moth hunters*. Canberra: Australian Institute of Aboriginal Studies.
Flood, J. 1983 *Archeology of the dreamtime*. Sydney: Collins.
Glatthaar, C., Welborn, T. A., Stenhouse, N. S. & Garcia-Webb, P. 1985 Diabetes and impaired glucose tolerance. A prevalence estimate based on the Busselton 1981 Survey. *Med. J. Aust.* **143**, 435–440.
Hiatt, L. R. & Jones, R. 1988 Aboriginal conceptions of nature. In *Australian science in the making* (ed. R. W. Home), pp. 1–22. Cambridge University Press in association with the Australian Academy of Science.
Jones, R. 1980 Hunters in the Australian coastal savannah. In *Human ecology in savannah environments* (ed. D. R. Harris), pp. 107–145. London: Academic Press.
Kirk, R. L. 1981 *Aboriginal Man adapting: the human biology of Australian Aborigines*. Melbourne: Clarendon Press.
Lowenstein, F. W. 1961 Blood pressure in relation to age and sex in the tropics and subtropics – a review of the literature and an investigation in two tribes of Brazil Indians. *Lancet* i, 389–392.
McArthur, M. 1960 Report of the Nutrition Unit. In *Records of the American–Australian scientific expedition to Arnhemland*, vol. 2 (*Anthropology and nutrition*) (ed. C. P. Mountford), pp. 1–143. Melbourne University Press.
Naughton, J. M., O'Dea, K. & Sinclair, A. J. 1986 Animal foods in traditional Aboriginal diets: polyunsaturated and low in fat. *Lipids* **21**, 684–690.
Neuringer, M., Connor, W. E., Van Petten, C. & Barstad, L. 1984 Dietary omega-3 fatty acid deficiency and visual loss in the infant Rhesus monkey. *J. clin. Invest.* **73**, 272–276.
O'Dea, K. 1984 Marked improvement in the carbohydrate and lipid metabolism in diabetic Australian Aborigines after temporary reversion to traditional lifestyle. *Diabetes* **33**, 596–603.
O'Dea, K. 1991 Westernization and non-insulin dependent diabetes in Australian Aborigines. *Ethnicity & Disease*. (In the press.)
O'Dea, K., Spargo, R. M. & Akerman, K. 1980 The effect of transition from traditional to urban lifestyle on the insulin secretory response in Australian Aborigines. *Diabetes Care* **3**, 31–37.
O'Dea, K., Spargo, R. M. & Nestel, P. J. 1982 Impact of westernization on carbohydrate lipid metabolism in Australian Aborigines. *Diabetologia* **22**, 148–153.
O'Dea, K. & Sinclair, A. J. 1985 The effects of low fat diets rich in arachidonic acid on the composition of plasma fatty acids and bleeding time in Australian Aborigines. *Int. J. Nutr. Vitaminol.* **33**, 441–453.
O'Dea, K., Naughton, J. M., Sinclair, A. J., Rabuco, L. & Smith, R. M. 1987 Lifestyle change and nutritional status in Kimberley Aborigines. *Aust. Aboriginal Stud.* **1**, 46–51.
O'Dea, K., Traianedes, K., Hopper, J. L. & Larkins, R. G. 1988*a* Impaired glucose tolerance, hyperinsulinemia and hypertriglyceridemia in Australian Aborigines from the desert. *Diabetes Care* **11**, 23–29.
O'Dea, K., White, N. G. & Sinclair, A. J. 1988*b* An investigation of nutrition-related risk factors in an isolated Aboriginal community in northern Australia: advantages of a traditionally orientated lifestyle. *Med. J. Aust.* **148**, 177–180.
O'Dea, K., Lion, R. J., Lee, A., Traianedes, K., Hopper, J. L. & Rae, C. 1990 Diabetes, hyperinsulinemia and hyperlipidemia in small Aboriginal community in Northern Australia. *Diabetes Care* **13**, 830–835.
Reaven, G. M. 1988 Role of insulin resistance in human disease. *Diabetes* **37**, 1595–1607.
Rutishauser, I. H. E. & McKay, H. 1986 Anthropometric status and body composition in Aboriginal women of the Kimberley region. *Med. J. Aust.* **144**, 8–10.
Sahlins, M. 1972 *Stone age economics*. Chicago: Aldine.
Sanders, T. A. B. 1985 Influence of fish-oil supplements on man. *Proc. Nutr. Soc.* **44**, 391–397.
Seidell, J. C., Deurenberg, J. G. & Hautvast, J. A. G. 1987 Obesity and fat distribution in relation to health – current insights and recommendations. *Wld Rev. Nutr. Diet.* **50**, 57–91.
Sinclair, A. J. & O'Dea, K. 1990 Fats in human diets through history: is the Western diet out of step? In Reducing fat in meat animals (ed. J. D. Wood & A. V. Fisher), pp. 1–47. London: Elsevier Applied Science.
Thorburn, A. W., Brand, J. C. & Truswell, A. S. 1987*a* Slowly digested and absorbed carbohydrate in traditional bushfoods: a protective factor against diabetes? *Am. J. clin. Nutr.* **45**, 98–106.
Thorburn, A. W., Brand, J. C., O'Dea, K., Spargo, R. M. & Truswell, A. S. 1987*b* Plasma glucose and insulin responses to starchy foods in Australian Aborigines: a population now at high risk for diabetes. *Am. J. clin. Nutr.* **46**, 282–285.
Tonkinson, R. 1978 *The Mardudjara Aborigines: living in the Dreamtime*. Sydney: Holt, Rinehart & Winston.
White, N. G. 1985 Sex differences in Australian Aboriginal subsistence: possible implications for the biology of hunter–gatherers. In *Human sexual dimorphism* (ed. J. Ghesquierre, R. D. Martin & F. Newcombe), pp. 323–361. London: Taylor & Francis.
White, N. G., Beck, W. & Fullager, R. 1988 Archeology from ethnography: the Aboriginal use of cycad as an example. In *Archeology with ethnography: an Australian perspective* (ed. B. Meehan & R. Jones). Canberra: Australian National University.
Wise, P. H., Edwards, F. M., Thomas, D. W., Eliott, R. B., Hatcher, L. & Craig, R. 1976 Diabetes and associated variables in the South Australian Aboriginal. *Aust. N.Z. J. Med.* **6**, 191–196.

Discussion

P. A. Jewell (*Physiological Laboratory, University of Cambridge, U.K.*). Periods of food abundance and food shortage occur. Were the shortages ever so extended as to threaten survival? Does Professor O'Dea have any observations that suggest that involuntary fasting was either detrimental or beneficial to the physical well-being of the people?

K. O'Dea. The available data since European contact indicate that severe food shortages did occur in association with prolonged droughts in arid areas of Australia (primarily inland). Food shortage was probably an important component of the process of establishing missions in such areas! There are a number of reports of Aboriginal groups coming into missions during droughts for food early in the colonization process.

Repeated cycles of involuntary fasting may have contributed to improved 'metabolic efficiency' of Aborigines. Recent studies in overweight people who diet regularly show that repeated cycles of weight loss and weight regain are associated with reduced energy requirements and linked with heightened preference for high-fat foods. Similar data have been reported in experimental animals. If such adaptation did occur, it would be expected to enhance the putative 'thrifty genotype' and could therefore have had important positive implications to survival of Aborigines as hunter–gatherers.

A. WHITEN (*Scottish Primate Research Group, University of St Andrews, U.K.*). We call the mode of foraging 'gathering and hunting', but what proportion of plant foods are in fact gathered and shared at a home base, and what proportion consumed while foraging, in the fashion of non-human primates? And what foods (and proportion of foods) are eaten without cooking?

K. O'DEA. Foods that require long preparation (e.g. winnowing and grinding of seeds, processing of cycad nuts, cooking of large animals) were usually taken back to the homebase and shared there. It is difficult to generalize with any accuracy about the proportion of foods that are either edible raw (fruits, some nuts) or require minimal cooking, that were consumed as they were gathered. Considerable amounts of food are eaten as they are gathered and major meals may also be consumed in this way: fish, small animals, shellfish and anything else that could be cooked quickly on a small fire (even the liver from a kangaroo) would frequently be eaten before returning to camp. However, some food was always carried back to the home base to share with others.

It has been estimated that in one region of northern Australia, more than half of the plant foods require cooking or some form of processing. However, this proportion would vary greatly both seasonally and with geographical location.

S. A. ALTMANN (*Department of Ecology and Evolution, University of Chicago, Illinois, U.S.A.*). What is known about the heritability of the physiological differences between Australian Aborigines and Europeans?

K. O'DEA. This is a difficult question to answer, mainly because of the problems of excluding potential confounding effects of environmental factors. Although it is widely believed that metabolic factors such as insulin resistance have important genetic components, there is no doubt that they are also influenced strongly by non-genetic factors (diet, lifestyle, etc.). Most of the data suggesting genetic differences in disease susceptibility (e.g. to diabetes) comes from cross-sectional comparisons of populations (Aboriginal, Caucasian) living under very different lifestyle conditions. The most that could be safely concluded at this stage is that proportionally more Aborigines appear to be susceptible to developing conditions such as obesity and NIDDM when living a 'western' lifestyle. Longitudinal family studies are essential to clarify this question. The evidence appears to be stronger with parameters such as body build (presumably largely genetically determined) where there do seem to be consistent differences between Aborigines and Caucasians, with Aborigines having a more 'linear' and lighter body build (relatively long limbs and lighter frame). However, it is possible that differences in nutritional status in early life may also influence these parameters.

S. S. STRICKLAND (*London School of Hygiene and Tropical Medicine, Keppel Street, WC1*). What is known about the lean and fat tissue composition of weight gain and loss over the feasting–fasting cycle?

K. O'DEA. There are no data on this question in Aborigines. In Caucasians there are data suggesting that repeated cycles of weight loss and weight regain in overweight individuals results in enhanced 'metabolic efficiency', possibly secondary to a reduction in the relative proportion of lean body mass. Studies in rats suggest that this enhanced metabolic efficiency associated with weight cycling can be prevented by exercise training. To my knowledge, data on the consequence of major fluctuations in energy intake on body composition in lean, physically active individuals are not available.

S. S. STRICKLAND. Professor O'Dea described the aborigines as 'lean', with a mean body mass index of 16.7, yet a BMI of 20, equivalent to the body fatness of a Caucasian with a BMI of 22–23. Can she comment on differences in body composition and the distribution of body fat between aborigines and caucasians?

K. O'DEA. There is a paucity of good quality data on body composition in Aborigines. However, what data there are indicate that Aborigines have a linear body build, with proportionally longer limbs and shorter torso than Caucasians. This type of body build is consistent with the data reported indicating that for an equivalent BMI young Aboriginal women have more body fat (from measurements of skinfold thickness and bioelectrical impedance) than young Caucasian women. There is now an accumulating body of evidence indicating that both male and female Aborigines develop a central, or android, pattern of fat distribution when they gain weight.

O. T. OFTEDAL (*Smithsonian Institution, Washington D.C., U.S.A.*). I am intrigued by Professor O'Dea's observation that the Aborigines had detailed knowledge of the seasonal cycles of fat content in their prey. Could she elaborate on this point with regard to types of prey that exhibit such cycles, and the efforts made to procure them?

K. O'DEA. Most types of prey have times of the year when they are fatter than others, with clearly larger fat depots within the abdomen, rather than increased fat content of the muscle. The seasonal movement of Aboriginal groups within their territory was closely related to the fatness of the major animal food sources. The importance of this 'fat-possessing quality' of foods was reflected in the complex language to describe it across the edible fauna: insects, fish, crustacea, molluscs, reptiles, birds and mammals.

Hunting income patterns among the Hadza: big game, common goods, foraging goals and the evolution of the human diet

K. HAWKES[1], J. F. O'CONNELL[1] AND N. G. BLURTON JONES[2]

[1] *Department of Anthropology, University of Utah, Salt Lake City, Utah 84112, U.S.A.*
[2] *Graduate School of Education, and Departments of Anthropology and Psychiatry, University of California, Los Angeles, California 90024, U.S.A.*

SUMMARY

The assumption that large mammal hunting and scavenging are economically advantageous to hominid foragers is examined in the light of data collected among the Hadza of northern Tanzania. Hadza hunters disregard small prey in favour of larger forms (mean adult mass $\geqslant$ 40 kg). Here we report experimental data showing that hunters would reduce their mean rates if they included small animals in the array they target. Still, daily variance in large animal hunting returns is high, and the risk of failure correspondingly great, significantly greater than that associated with small game hunting and trapping. Sharing large kills reduces the risk of meatless days for big game hunters, and obviates the problem of storing large amounts of meat. It may be unavoidable if large carcasses cannot be defended economically against the demands of other consumers. If so, then large prey are common goods. A hunter may gain no consumption advantage from his own big game acquisition efforts. We use Hadza data to model this 'collective action' problem, and find that an exclusive focus on large game with extensive sharing is not the optimal strategy for hunters concerned with maximizing their own chances of eating meat. Other explanations for the emergence and persistence of this practice must be considered.

1. INTRODUCTION

Many accounts of the early stages of human evolution assume that the practice of hunting and scavenging large animals arose and persisted because of the nutritional advantages hominid hunters earned for themselves and their families (Isaac 1978). The same assumption underlies conventional explanations for changes in human foraging strategies observed in many parts of the world at the end of the Pleistocene: widespread declines in the abundance of large game, previously favoured for nutritional reasons, forced a general increase in diet breadth, commonly involving increased use of small game and high-cost plant foods, and culminating in some places in the development of agriculture (Cohen 1989). Although argument about the timing of the first transition has recently been quite vigorous (Binford 1984), the underlying rationale for both remains unchallenged, apparently because it seems so self-evident.

Here we report data on hunting among the Hadza of northern Tanzania. The Hadza are especially interesting because, unlike other low-latitude foragers operating without firearms or dogs, they take big game to the virtual exclusion of small-bodied prey. Their hunting and scavenging incomes may reflect in the clearest way features associated with big game specialization. We report results of an experiment designed to measure the income Hadza hunters could earn if they took small game. These data, combined with our observations of big game hunting, show that specialization in large animals maximizes mean daily rates measured in kilograms of meat. However, success rates for big game hunting, measured as the chance of acquiring a carcass on any day, are very low. The high average returns come at the expense of extreme variance. Big game hunting is a risky strategy in this habitat.

Sharing large carcasses is an effective way to reduce the risk. It is common among the Hadza, widely reported among other hunter–gatherers (Sahlins 1972), and generally considered to have a more or less ancient origin, crucial to human evolution (Isaac 1978; Lovejoy 1981; Binford 1984). Most anthropologists take it to be the likely outcome of an emphasis on big game hunting, arguing that everyone involved does better as a result.

The sharing of large carcasses solves two problems. It reduces the risk of meatless days for a hunter and his family, and obviates the problem of storing large amounts of meat. This consequence, however, may not explain the evolution and persistence of big game hunting and sharing. Among the Hadza, as among other hunter–gatherers (see, for example, Marshall 1961), meat from a large animal is taken and consumed by all, whether or not they paid the costs of capture.

Large carcasses are too big to defend against the demand for shares from other consumers. If sharing large game results from the economics of defence, these carcasses can be seen as 'common goods'. Given this, the evolutionary question is not why they are shared, but why anyone provides them in the first place. Ignoring the alternative of foraging for other resources, and assuming that animal tissue has special value (Hill 1989), a hunter's choice is between: (i) hunting and scavenging large game, which he will acquire at low frequency, and of which he will retain only a small portion; and (ii) hunting or trapping small animals, which he will take more often, and keep for himself and his family. We model this decision, using the Hadza data, to identify the evolutionarily stable strategy (ESS) for maximizing the number of days on which the hunters' children have meat to eat. The model shows that, although mean daily meat consumption would be much less, a hunter attempting to achieve this goal should take small game.

Clearly, hunters took big game in the past and continue to do so today. Among the Hadza, as among other well known modern hunter–gatherers, big game hunting is deemed to be the important activity of adult men. We infer that nutritional benefits may not account for the evolution and persistence of this practice. As for common or collective goods in general, 'selective incentives', benefits other than consumption of the goods themselves, may be necessary to motivate individuals to provide them (Olson 1965; Hawkes 1990, 1991). Like other cases in which individuals pursuing their own self-interest fail to serve the common good (Hardin 1968), hunters aiming to maximize their chances of eating meat would then be eating less.

2. THE EASTERN HADZA

The Eastern Hadza are a population of about 750 subsistence foragers who live in the rugged hill country south and east of Lake Eyasi. The climate of this region is warm and dry; the vegetation primarily savannah woodland. At the time of European contact, around the beginning of this century, the Hadza may have had this country largely to themselves. Since then (and especially over the past 30 years), they have been subjected to a series of government and church-sponsored settlement schemes intended to encourage them to abandon the foraging life in favour of full-time farming. None of these have been successful; in each instance, some Hadza have managed to avoid settlement entirely and continued to live as subsistence hunters. The most recent scheme, initiated in 1988, was in operation while the experiment described below was carried out.

3. HADZA HUNTING

Data on Hadza hunting were collected on 256 days of residence in Hadza camps during the period 1985–1989. Systematic observations enabled us to monitor the presence and absence of residents, and interviews helped establish the range of activities performed away from camp. 'Focal person follows' provided quantitative data on the latter. Residents allowed us to identify and weigh all food brought back to camp, as well as food accumulated at various points during focal follows (see Blurton Jones *et al.* (1989); Hawkes *et al.* (1989); O'Connell *et al.* (1988*a b*, 1990) for details).

Hadza hunting takes two forms, intercept and encounter. Intercept hunting is practised only in the late dry season (August–October). Men build blinds overlooking water sources and along heavily used game trails, sit in them overnight and shoot large animals with poisoned arrows as they pass within range. Since Hadza men are always armed, encounter hunting is effectively in progress most of the time they are away from camp. Hadza also scavenge large animals killed by other predators (O'Connell *et al.* 1988*b*).

4. HADZA HUNTING SUCCESS

Over the course of the study period, the Hadza hunters with whom we were living killed or scavenged 72 large animals, with an average of one animal every 3.6 days of observation. Species most commonly taken were giraffe (*Giraffa camelopardalis*), zebra (*Equus burchelli*), impala (*Aepyceros melampus*) and warthog (*Phaecochoerus aethiopicus*). Table 1 shows the approximate number of hunters in the study camps, and the number of animals taken by intercept and encounter hunting and scavenging respectively. Table 2 shows the return rates gained per hunter–day overall, per hunter–day for encounter hunting and scavenging only, and per hunter–night for intercept hunting.

Return rates by mass are high. Overall, the Hadza took one large animal every 29 hunter–days, for an estimated return of 4.9 kg (live mass) per hunter–day. Intercept hunting was the most productive technique, yielding one animal every 18 hunter–nights, or about 7.5 kg per hunter–night. Encounter hunting and scavenging produced, on average, one animal every 45 hunter–days overall, one every 53 days in the late dry season, and one every 37 in the wet.

Small animals were rarely taken by Hadza hunters during the period covered by our fieldwork, and returns measured by mass were extremely low. When caught, they were usually consumed before the hunter returned to camp. This means that our record of small animal captures comes from focal follows of foraging men. During 45 follows in 1985–1986, on which a total of 75 hunter–days were monitored, men shot at small game (mainly guinea fowl (*Guttera* spp.) and francolin (*Francolinus* spp.)) about once or twice a day, but actually took only 14 individuals. Ten of these were immature hornbills snatched from the nest as the hunter walked by. The total mass of all small prey taken on these trips was about 4.65 kg, approximately 0.062 kg per hunter–day, nearly two orders of magnitude less edible tissue than the mean obtained from hunting and scavenging large animals (table 2). This implies that small game accounts for only about 1 % of the animal tissue taken by the average adult hunter.

This pattern of prey selection differs sharply from that of other low-latitude hunter–gatherers operating

Table 1. *Kills by hunters in study camps, grouped by season and method of acquisition (256 days of observation, 1985–1989)*

(Encounter hunting and scavenging are daytime activities practised all year long. Intercept hunting is a night-time activity practised only in the late dry season.)

season	year	observation days	approximate number of hunters per camp	encounter animals	encounter total/kg	scavenge animals	scavenge total/kg	intercept animals	intercept total/kg
late dry	1985	47	10	6	540	6	552	18	2465
wet	1985–1986	61	6	6	625	2	82	—	—
early dry	1986	36	6	3	440	2	1200	—	—
late dry	1986	44	10	4	150	2	48	7	877
late dry	1988	43	10	5	390	3	880	1	200
wet	1989	25	6	6	1666	—	—	—	—

Table 2. *Return rates in kilograms and numbers of prey, grouped by season and method*

	hunter–days	mean kg per hunter–day	s.d.	mean individual per hunter–day	s.d.	mean kg h^{-1}[a]
all methods daily rate	2072	4.890	39.736	0.034	0.182	—
encounter/scavenger all seasons	2072	3.181	33.325	0.022	0.146	0.71
late dry[b] encounter/scavenger only	1340	1.923	21.080	0.019	0.138	0.45
late dry[b] night (intercept)	473	7.488	45.383	0.055	0.228	—
wet	516	4.599	41.999	0.027	0.163	1.02
late dry all methods	1340	4.566	34.249	0.039	0.193	—

[a] For a sample of 80 observation days covering late dry and wet seasons in 1985–1986, mean number of hours spent by adult men in day-time foraging was about 4.5 hours (both seasons). We use this number to calculate an hourly rate.

[b] In the late dry season of 1985 hunters spent, on average, every third night intercept hunting from blinds. We use these data to calculate rates for intercept hunting overall. For daytime late dry season rates, we exclude intercept kills.

on foot without dogs or firearms, many of whom take small animals often. For example, Lee's (1979) work diaries on !Kung hunters in the Dobe area show that, during a four week study period, all prey taxa taken without the use of dogs (an estimated 29% of total animal prey by mass) had adult body mass less than 20 kg. Yellen (1977), reporting data from the same population, shows that over several months, about 55% of total prey mass taken was derived from small game. For the Ache of Eastern Paraguay, Hill & Hawkes (1983) report that small animals account for more than 75% of prey mass acquired by bow hunters.

5. THE SMALL GAME HUNTING EXPERIMENT

The conventional wisdom on hunter–gatherer diets leads us to expect that the Hadza hunters specialize in taking large prey because they do better nutritionally than they would by taking smaller animals. We can test this hypothesis by means of the optimal diet model (Charnov 1976), which is designed to predict which resources a forager should select from among an available array (and which it should ignore), given the goal of maximizing mean nutrient acquisition rate. If we hypothesize that Hadza hunters seek this goal, the model helps us identify the suite of prey that best meets it. The model assumes random encounter with prey and knowledge of probable encounter rates with available prey types. Whether any particular prey item falls in the optimal set, that which maximizes mean rates, depends on the net benefits associated with pursuing it, against ignoring it in favour of other resources. Our data suggest that, during the wet season, a Hadza hunter can expect to earn a long term average of about 4.6 kg per day, or about 1 kg h^{-1}, from encounter hunting and scavenging large animals. If his goal is to maximize that rate, he should ignore any potential prey item likely to yield a lower return. If the conventional wisdom on hominid prey choice and its determinants is correct, small animals should be among those resources yielding lower average returns. If, however, men hunt big game for other, non-nutritional reasons, we might find that small game hunting yields higher average returns than does the pursuit of larger prey.

During October–November 1990, we enlisted the assistance of several Hadza men, then living in a settlement, who agreed to accompany us back to the bush and participate in an experiment designed to provide quantitative data on foraging returns available from targeting small game to the exclusion of other animal prey. All the men had lived as subsistence hunters for most of their lives; all had been among the subjects of previous research on foraging in precisely the area in which the experiment was carried out. In return for their participation, all were provided daily rations and a wage. An observer followed a different hunter, or pair of hunters, each day. Men not followed presented all prey taken to be tallied and weighed. This produced two data sets, the smaller one consisting

Table 3. *Prey species pursued by focal men during small game hunting experiment and return rates post-encounter for those frequently pursued. Totals for 'all days' include prey taken by non-focal men*

prey species	number of pursuits	minutes in pursuit	kg taken on focals	kg all days	individual prey all days	kg h^{-1} after encounter
guinea fowl (*Guttera* sp.) or francolin (*Francolinus* sp.)	81	131	0.50	7.13	9	0.23 (0.90)[a]
hyrax (*Procavia capensis*)	52	819	9.52	11.57	9	0.66
dikdik (*Rhynchotragus* sp.)	21	35	0.90	0.90	1	< 1.54 (0.42)[b]
baboon (*Papio anubis*)	1	5	—	6.00	2	
tortoise	1	3	0.60	2.80[c]	3[c]	
impala (*Aepyceros melampus*)	14	26				
bird	6	4				
zebra (*Equus burchelli*)	2	13				
bat eared fox (*Otocyon megalitis*)	1	8				
buffalo (*Syncerus caffer*)	1	4				
giraffe (*Giraffa camelopardalis*)	1	4				
python	1	1				
kudu (*Tragelaphus strepsiceros*)	1	1				
small mammal	2	6				

[a] Only one of the pursuits during focal follows was successful. We use the prey captured over all days, and assume the same pursuit times per hunter–day for non-focal hunters to estimate a total pursuit time, and so estimate an over-all post-encounter rate (in parentheses).

[b] As only one (immature) dikdik was taken by encounter hunting during the experiment, (and that one during a focal follow) we consider the measured focal rate to be an extreme value. An overall post-encounter rate estimated as in [a] above is given in parentheses. It also depends on a single capture and may be high.

[c] A 2.2 kg tortoise was brought to camp, then released, hunter unknown. This was not counted in tallying rates.

of detailed time allocation records for focal men, the larger of daily hunting incomes for all men.

6. RESULTS

The experimental period ran from 17 October–6 November 1990. Heavy rains, marking the beginning of the wet season, fell intermittently throughout. Data were collected on seven men over 16 days, for a total of 102 man–days. Income from hunting small prey averaged 0.225 ± 0.480 small animals per hunter–day, or about 0.252 ± 0.626 kg per hunter–day. Prey taken included francolin, guinea fowl, hyrax (*Procavia capensis*), dikdik (*Rhychotragus* spp.), and baboon (*Papio anubis*) (see table 3).

Follows were conducted on 15 days. As men usually foraged in pairs, follows produced detailed records for 28 focal man–days. Most days included two foraging bouts, separated by a noon-time rest in camp. Focal men spent an average of 411 minutes per day foraging. Although each participant was asked to take as great a mass of small animals as possible, focal men spent time in other activities, mainly honey collecting. A daily average of 41.4 ± 29.6 min was invested in this, yielding a total of 21.89 kg of honey over 33 collecting events, for a mean daily rate of 0.78 kg per man–day. Focal men also pursued larger game on 19 occasions, never successfully.

Men were encouraged each day to apply themselves seriously to the assigned task. Focal men were actively encouraged to start hunting in the morning and after any mid-day break. The vigilant eye of the observer may have had an effect. Daily income was higher, although not significantly, for focal against non-focal men: 0.408 ± 0.750 kg per day and 0.429 ± 0.790 animals per day ($n = 28$), against 0.200 ± 0.593 kg per day and 0.162 ± 0.362 animals per day ($n = 74$) (Mann–Whitney U-test, $p = 0.073$ for income measured by mass, $p = 0.140$ for number of animals).

Both small game encounter rates and return rates from subsequent pursuit can be estimated on the basis of data collected during the focal–person follows. Pursuits on 17 different species were observed, but only four (guinea fowl, francolin, hyrax and dikdik) were encountered often enough to provide useful samples (table 3). However, the number of encounters is not necessarily a measure of abundance. Some prey types may be so cryptic or elusive that actual encounters (defined as visual contacts in which pursuit was initiated) are low. Dikdik, for example, are seen many times in the course of an average day's foraging, but are so quick to flee that the encounter rate, marked by the fact that the hunter raised his bow, is only about one per day.

Encounters and return rates per small prey taxon are listed in table 3. Because most pursuits are unsuccessful, post-encounter return rates are low; for three of the four taxa for which useful samples are available, return rates are clearly lower (0.23–0.66 kg per hour) than the long term one kg per hour average for large mammal encounter hunting and scavenging in the wet season. We cannot reject the hypothesis that returns for the fourth taxon, dikdik, are also less than one kg per hour. According to the optimal diet model, a Hadza hunter does indeed maximize his average rate of meat acquisition by generally ignoring these taxa in favour of larger prey, at least in this season.

(*a*) *Snaring*

During the first day of the experimental period, hunters suggested that guinea fowl and dikdik could be taken more effectively with snares. They were encouraged to pursue this or any other traditional strategy they thought might be more productive. None of them did so. Instead, they directed the two young men (later one) who accompanied the party as camp assistants to set and monitor snares.

About 16 snares were in operation for the duration of the experiment. All made use of a bent sapling, twine loop and a simple trigger. Twine for all snares was made by one of the hunters who participated in the experiment. It took about 15–20 minutes to set each snare the first time, but only a minute or so to reset it once it was sprung. Time spent on snares on any given day was always much less than the time focal men spent in encounter hunting. Attention paid to snares was inconstant. The young men assigned to monitor the snares were not themselves regularly monitored. On two occasions, hunters reported that dikdiks had been snared and were then stolen by leopards. Birds were also lost from the snares to other consumers. On one day, a guinea fowl foot, the only part remaining in the snare, was presented for weighing. On several occasions, an observer passed snares which were not set.

Two points should be underlined. First, the observed return rate for snares is unlikely to be the maximum that could be earned from this technique in this season in Hadza country. Second, snares in this environment require attentive monitoring because of wind and the density of competing predators. Unless they are actively watched, they may hang long unset or serve as a bonanza for others.

Seventeen individual prey (ten francolin, two guinea fowl, one dikdik and four small unidentified birds) were taken with snares, at least one on eight of the 14 days the experiment lasted. Daily income averaged 1.429 ± 2.174 animals per day, or about 0.781 ± 1.097 kg per day ($n = 14$). These figures are significantly higher than the average returns from small animal encounter hunting (Mann–Whitney U-test for daily income by number of animals, $p = 0.0008$; by mass, $p = 0.0029$). Income by mass from trapping is significantly lower than the long-term mean available from large mammal encounter hunting and scavenging in this season. Mean rate maximizers would therefore choose big game hunting and scavenging rather than snaring.

7. DISCUSSION

These data show that Hadza hunters who encounter hunt instead of monitoring snares, and who ignore small prey in favour of continued search for large prey, increase their long-term mean rate of acquiring meat. In doing so, they also increase their risks of failing to get anything at all. As our small game experiment was conducted in the wet season, we pay most attention to the wet-season data. The extreme risk of failure for big game hunters can be expressed in several ways. Hadza men take only 0.027 ± 0.305 individual large carcasses per hunter–day in the wet season, approximately one large animal every five weeks (37 days). The chance of failure on any given day is just over 97%. In these circumstances, a man relying on his own work to get food for himself (let alone any dependants), and choosing big game hunting and scavenging as his procurement strategy, clearly has a very risky future.

By comparison, the daily success rate for !Kung hunters is 0.23 (about one animal every four man–days, on average; Lee 1979: 267); for Ache men it is 0.57–0.76 (2–3 successful days out of every four; Hill & Hawkes 1983: 176; Hawkes 1990: 151, respectively, using different samples). These hunters have a 'failure rate' that is an order of magnitude lower, largely because they include small game in the suite of prey they exploit. The Ache, and perhaps the !Kung, maximize their mean rate of meat acquisition by including small prey. The size and local abundance of large prey enable Hadza hunters to maximize their mean rates by specializing in big game; but in doing so, they accept an extremely high risk of failure.

Sharing can ameliorate this risk when large food packages are acquired asynchronously (Kaplan & Hill 1985; Cashdan 1985; Winterhalder 1986). The Hadza represent just such a case. If we assume that when any Hadza hunter kills a large animal every member of his residential group eats part of it, the probability of eating meat on any given day (in the wet season) is $1-(0.973)^f$, where f is the number of active hunters. If that number is six (the typical figure for wet season study camps), the chance of eating meat from a large carcass is 0.15, (i.e. meat once a week). If meat from a single carcass may be available for more than one day (a reasonable assumption for animals the size of zebra (mean adult mass 200 kg) or larger), consumption frequency is increased. If meat lasts three days after a kill, and all eat on those three days, then the overall probability of eating meat on any given day is $1-(0.973)^{3f}$. Where $f = 6$, this is 0.39. These are better odds.

In these circumstances, where all hunters specialize in big game and earn a daily average return of 4.6 kg (the observed wet season mean for the study period), the sharing pattern would give the members of each hunter's household (assuming there are six) an average of 2.3 kg every third day on average. If any hunter and his household left the big game hunting and sharing group, and the hunter himself elected to target small game, he would probably take about one animal every other day, for a daily average return of 0.4 kg. Members of his household would receive 0.13 kg every other day. If he adopted a trapping strategy, his dependants would eat about twice as much meat, although only about one fifth as much as did the dependants of a big game hunter.

Big game hunting and sharing provides more meat for everyone, just as the conventional wisdom would have it. The variance for the big game hunters would be higher, which might have more deleterious effects on children than adults. This potentially important problem aside, all would do better if each hunter maximized his mean rate of meat acquisition and specialized in big game. Still, the question of how these practices could evolve and persist turns not on their consequence for the group, but on their consequences

Table 4. *Pay-off matrix*

(P = the probability of failure on any day for a big game hunter, f = the number of foragers, SG = the probability of not failing for a small game hunter, SN = the probability of not failing for a snarer.)

	others hunt or scavenge big game	others hunt small game	others snare
hunt or scavenge big game	$A = 1-(P)^{3f}$	$B = 1-(P)^3$	$C = 1-(P)^3$
hunt small game	$D = [1-(P)^{3(f-1)}]+SG$	$E = SG$	$F = SG$
snare	$G = [1-(P)^{3(f-1)}]+SN$	$H = SN$	$I = SN$

for each individual hunter and the pay-off he could get from other choices.

8. THE COLLECTIVE ACTION PROBLEM

The advantages of sharing among specialized big game hunters hinge on what any hunter claims from the kills of others. A Hadza hunter earning the average 4.6 kg per day in the wet season and, sharing with five other hunters and their families, keeps for himself and his family only one sixth of his own income, 0.77 kg per day. If we count only what he and his family consume as his actual foraging income, he earns about what he would get by snaring, but with seven times the variance. In addition to what he brings in for his family, his big game hunting provides an equivalent amount for each co-resident hunter. As each of them does the same for him, it is the big game hunting of others that provides most (in this case five sixths) of what he and his family consume. This might be viewed as reciprocity; the shares he gives to others may be the price he pays for receiving in turn from them. The assumption that current contributions are the price of future benefits returned is central to models of reciprocal altruism (Trivers 1971). Some have argued that such contingent cooperation is extremely general (Axelrod 1984; but see Boyd & Lorberbaum 1987; Boyd & Richerson 1988; Hirshleifer & Martinez-Coll 1988).

Ethnographic characterizations of hunter–gatherer societies undercut this generalization, often emphasizing the non-contingent character of food sharing (Sahlins 1972), especially of meat (Kaplan & Hill 1985; Marshall 1961). Shares are distributed irrespective of contributions to the pot. Active hunters can neither sanction slackers nor refuse scroungers. In cases like the Hadza, high variance and skewed income distributions associated with large-animal procurement create additional problems of accounting. Large samples and long periods of monitoring would be required to distinguish between hunters who were out of camp but not investing time in hunting and scavenging big game, and those who were regularly seeking large animals but were either inept or unlucky.

Many forms of cooperation cannot be taken by the recipient, but must be initiated by the donor. Food is different, and its defence may be costly (Blurton Jones 1984, 1987). Observations among the Hadza suggest that small game are easily hidden and protected from the demands of others. With larger prey, defence becomes expensive. One way to view the sharing of large carcasses is to see them as common or collective goods, i.e., goods which, like 'public goods', can be consumed by those who do not pay for them (Olson 1965; Hardin 1982). From this perspective, the question of why they are shared is not problematic: it follows from the prohibitive costs of exclusion. The question instead is whether a hunter in a group, knowing that most of any large carcass he captures will be consumed by others, finds it in his interest to specialize in acquiring them in the first place. If he does, he earns as much for others as for himself. Would he do better to take advantage of whatever other hunters earn from specializing on big game, but devote his own efforts to small animals for himself and his family?

We can model this decision as a game (Maynard Smith 1982) in which one forager chooses between big game hunting, small game hunting and snaring in a 'field' of other foragers; and evaluate whether any of these are stable strategies, given that the pay-offs are frequency dependent (i.e. one forager's chance of eating meat depends on what the others do). Assume that if any forager hunts or snares small game and is successful, he (and only he) eats his catch on that day. To include the assumption that a big game kill provides a large enough amount to cover several days, say three, assume: (i) that foragers cannot starve or suffer nutritional depletion for lack of animal tissue over the first two days; (ii) that they commit to a strategy for three days in sequence; and (iii) the crucial day of reckoning is the third.

Which of these strategies will be the best choice for a forager trying to maximize his chances of not starving for meat? In the matrix shown in table 4, the focal forager's choices are represented by the rows, and his pay-offs (his chances of getting meat) appear in the cells of the matrix. The columns are the strategies of the 'field' (the other foragers in the group). The simple assumption here, that members of the field all adopt the same strategy, allows us to see which strategy, if any is 'evolutionarily stable', i.e. once established, cannot be bettered by an alternative.

During the Hadza experiment focal hunters took small animals on nine of 28 days. The minimum take on a successful day was 0.5 kg. Assuming that 0.5 kg is a day's ration of meat, the chance of eating by hunting small game on any day (the value for SG in the matrix) is $\frac{9}{28}$, or 0.32. Snaring was successful on eight of 14 days, but yielded more than 0.5 kg on only seven of those days. If less than 0.5 kg is not enough, the chance of getting enough meat to eat on any snaring day (the value for SN in the matrix) is $\frac{7}{14}$, or 0.50. Given that the total number of hunters is six, the other values are:

$A = 0.39$, $B = C = 0.08$, $D = 0.69$, $E = F = 0.32$, $G = 0.87$, $H = I = 0.50$. Any hunter does better if others hunt big game: his highest pay-offs are all in the first column. But whatever the others do, he maximizes his chances of getting meat if he snares: the highest values in each column are all in the bottom row. If we assume snaring is not an option, and remove the last row and column, the hunter's best chance of eating meat requires taking small prey. If we use the overall rates for the small game experiment, rather than those for the focal hunters, the values of A and B remain the same, but those of D and E are lower. Incomes were greater than 0.5 kg on 19 of 102 hunter–days. The small game success rate is 0.19; thus $D = 0.56$, $E = 0.19$. Whether the others hunt big game or not, a forager increases his chances of eating meat by going for small animals.

This pay-off structure has the form familiar to game theorists and students of the evolution of cooperation as a 'prisoner's dilemma' (for example, see Hardin 1982). In some circumstances, cooperation can be evolutionarily stable in iterated prisoner's dilemmas (see, for example, Boyd 1989), but not so here. Strategies of cooperation which can persist are always contingent; they involve withholding benefits from those who fail to cooperate. Here, where the cooperation is the provision of common goods, individuals who do not themselves contribute cannot be prevented from consuming what has been provided by others. In these circumstances 'free riders' thrive, and a 'tragedy of the commons' follows (Hardin 1968). Although all would do better if only each contributed to the common or collective good, each does better for himself by maximizing his private benefits.

9. GENERALITY OF RESULTS

Our results are a function of values measured or calculated for the Hadza during the periods covered by our observations. Are they more broadly applicable, or do they depend on special features of this particular case? The small game rates from our experiment are very low, but results from the model are sufficiently robust that they would hold, even if lower still. Increasing them would further increase the relative advantages of small game hunting. The results depend most strongly on our assumption that large game are shared and small game are not. Given this, acquiring food for the consumption of the hunter or his offspring cannot be the adaptive function of big game specialization. Hunters pursuing that goal should take small game.

This does not mean that a forager should ignore opportunities to take large animals whenever they arise. Taking large carcasses when chance presented them would have little effect on the failure rates of small game hunters, in that chances to acquire large carcasses would necessarily be even rarer for small game hunters than for big game specialists. Occasional large carcasses would raise the long-term mean return rates of small game hunters; but they would still earn lower long-term means than they could by ignoring small animals. The resulting diet would include big game, but in considerably smaller numbers than we see among the Hadza. In the circumstances we have described, hunters seeking to feed themselves and their families should not be mean-rate maximizers.

Just as the small game hunting and trapping rates we estimate for the Hadza may be unusually low, so the big game return rates may be high, especially in comparison with the prehistoric past. The poisoned, metal-tipped arrows used by Hadza men should increase their efficiency as competitive scavengers and big game hunters relative to that of hunters who lack this technology. Adjusting Hadza big game rates to compensate for this technology would make big game specialization even less successful at earning daily consumption requirements than we have calculated.

However, populations of large herbivores and their carnivore predators may be smaller in Hadza territory than elsewhere, especially in the past. Although the Hadza habitat has been characterized as unusually rich by comparison with the habitats of other modern hunter–gatherers, local game densities are lower than those in at least some other modern East African habitats (O'Connell *et al.* 1988*b*), and probably lower than those in many parts of the world during the marked fluctuations of the Pleistocene. If populations of large herbivores and their predators were more dense, big game hunting and scavenging rates might rise accordingly.

This caveat aside, we interpret our results to support scepticism that nutritional advantages to hunters and their families can account for the persistence of specialized big game hunting. We conclude that explanations for the practice here, and in different times and places where hunting incomes have similar patterns, may require us to investigate other benefits that serve as the 'selective incentives' to make big game hunting pay. As it pays for men but not women, mating advantages seem likely candidates.

This work was financed by the National Science Foundation, the Swan Fund, Ms B. Bancroft, the University of Utah, and the University of California at Los Angeles. We thank Utafiti (Tanzanian National Research Council) for permission to pursue fieldwork; we thank D. Bygott and J. Hanby for continued vital assistance, and the Hadza themselves for tolerance, advice and support. C. O'Brien, District Game Officer, A. Shanny, P. Shanalingigwa and associates provided assistance for the small game experiment. E. Cashdan, K. Hill, P. Lee, A. Rogers, A. Whiten and D. Zeanah provided useful comments on earlier drafts.

REFERENCES

Axelrod, D. 1984 *The evolution of cooperation.* New York: Basic Books.

Binford, L. 1984 *Faunal remains from Klasies River mouth.* New York: Academic Press.

Blurton Jones, N. 1984 A selfish origin for human food sharing: tolerated theft. *Ethol. Sociobiol.* **5**, 1–3.

Blurton Jones, N. 1987 Tolerated theft: suggestions about the ecology and evolution of sharing. *Social Sci. Inf.* **326**, 31–54.

Blurton Jones, N., Hawkes, K. & O'Connell, J. 1989 Studying costs of children in two foraging societies: implications for schedules of reproduction. In *Comparative*

socioecology of mammals and man (ed. V. Standon & R. Foley) pp. 365–390. London: Blackwell.
Boyd, R. 1989 Prisoner's dilemma game. *J. theor. Biol.* **136**, 47–56.
Boyd, R. & Lorberbaum, J. 1987 No pure strategy is evolutionarily stable in the repeated prisoner's dilemma game. *Nature, Lond.* **327**, 58–59.
Boyd, R. & Richerson, P. 1988 The evolution of reciprocity in sizeable groups. *J. theor. Biol.* **132**, 337–356.
Cashdan, E. 1985 Coping with risk: reciprocity among the Basarwa of northern Botswana. *Man* **20**, 454–474.
Charnov, E. 1976 Optimal foraging: the attack strategy of a mantid. *Am. Nat.* **110**, 141–151.
Cohen, M. 1989 *Health and the rise of civilization*. New Haven: Yale University Press.
Hardin, G. 1968 The tragedy of the commons. *Science, Wash.* **162**, 1243–1248.
Hardin, R. 1982 *Collective action*. Baltimore: The Johns Hopkins University Press.
Hawkes, K. 1990 Why do men hunt? Benefits for risky choices. In *Risk and uncertainty in tribal and peasant economies* (ed. E. Cashdan), pp. 145–166. Boulder: Westview Press.
Hawkes, K. 1991 Showing off: tests of an hypothesis about men's foraging goals. *Ethol. Sociobiol.* **12**, 29–54.
Hawkes, K., O'Connell, J. & Blurton Jones, N. 1989 Hardworking Hadza grandmothers. In *Comparative socio-ecology of mammals and man* (ed. V. Standen & R. Foley), pp. 341–366. London: Blackwell.
Hill, K. 1989 Macronutrient modifications of optimal foraging theory: an approach using indifference curves applied to some modern foragers. *Hum. Ecol.* **16**, 157–197.
Hill, K. & Hawkes, K. 1983 Neotropical hunting among the Ache of eastern Paraguay. In *Adaptations of native Amazonians* (ed. R. Hames & W. Vickers), pp. 139–188. New York: Academic Press.
Hirshleifer, J. & Martinez-Coll, J. 1988 What strategies can support the evolutionary emergence of cooperation? *J. Conflict Resol.* **32**, 367–398.
Isaac, G. 1978 The food sharing behaviour of proto-human hominids. *Scient. Am.* **238**, (4), 90–108.
Kaplan, H. & Hill, K. 1985 Food sharing among Ache foragers: tests of explanatory hypotheses. *Curr. Anthropol.* **26**, 223–245.
Lee, R. 1979 *The !Kung San: men, women, and work in a foraging society*. Cambridge University Press.
Lovejoy, C. O. 1981 The origin of man: a review. *Science, Wash.* **211**, 341–350.
Marshall, L. 1961 Sharing, talking and giving: relief of social tensions among !Kung Bushmen. *Africa* **31**, 231–249.
Maynard Smith, J. 1982 *Evolution and the theory of games*. Cambridge University Press.
O'Connell, J. F., Hawkes, K. & Blurton Jones, N. 1988*a* Hadza hunting, butchering, and bone transport and their archaeological implications. *J. anthrop. Res.* **44**, 113–162.
O'Connell, J. F., Hawkes, K. & Blurton Jones, N. 1988*b* Hadza scavenging: implications for Plio-Pleistocene hominid subsistence. *Curr. Anthropol.* **29**, 356–363.
O'Connell, J. F., Hawkes, K. & Blurton Jones, N. 1990 Reanalyses of large mammal body part transport among the Hadza. *J. archaeol. Sci.* **17**, 301–316.
Olson, M. 1965 *The logic of collective action*. Cambridge, Massachusetts: Harvard University Press.
Sahlins, M. 1972 *Stone Age economics*. Chicago: Aldine.
Trivers, R. 1971 The evolution of reciprocal altruism. *Q. Rev. Biol.* **46**, 35–57.
Winterhalder, B. 1986 Diet choice, risk and food sharing in a stochastic environment. *J. anthrop. Res.* **5**, 369–392.
Yellen, J. 1977 *Archeological approaches to the present: models for reconstructing the past*. New York: Academic Press.

Discussion

O. T. Oftedal (*Smithsonian Institution, Washington D.C., U.S.A.*). When the authors refer to kilograms of prey obtained, do they know or can they estimate the yield of edible flesh? Also it is important that the fat content of the edible portion be measured, if possible, as that will greatly influence the energetic return from hunting. Both factors may have a bearing on the relative advantages of large against small prey.

K. Hawkes. Yes they may, The numbers we used here are carcass masses of prey which include inedible fractions like bone and which ignore differences in fat and protein content. The nutritional value of what hunters earn depends on both the edible portions and their nutrient composition. These variables have not yet been measured for Hadza prey. Work is currently underway by our research group (O'Connell and Lupo) to do so, stimulated especially by systematic differences across species in the treatment at kill sites of carcasses in the same size class (e.g. zebra against alcelaphine antelope) which may be due to differences in body composition. The ranges measured and estimated for edible fractions of game animals elsewhere vary widely. Lee (1979) estimates this fraction for game taken by !Kung hunters in northern Botswana to be 50%. Hart (1978) calculated edible portions over 80% for game taken by Mbuti Pygmies in the Ituri Forest of Zaire. A few measurements of prey taken by Ache hunters in eastern Paraguay range between 69% and 88% edible (Hill *et al.* 1984). Body fat estimates vary widely by species, sex and season, and our experience suggests marked individual variation as well. If there is a systematic difference in edible fraction and fat composition by body size alone, it could affect the foraging choices discussed here. Hadza small prey may show the most extreme variance in these dimensions. Birds may be generally higher in edible fraction than mammals and lower in fat content. Hyraxes look like round packages of fat. I expect they have much higher fat fractions than the large ungulates. The big game animals are lean enough (Ledger 1968) that seasonal differences (and individual variation within season) of a few percent in fat can make a very large difference in nutritional value per edible kilogram. All of the potential sources of variability present opportunities for error in turning prey weights into nutritional values. More precise cost–benefit estimates would also include processing costs which have been ignored here. These costs might vary by prey size; for example, there may be some 'economy of scale' in that the cost per kilogram of skinning, butchering and cooking a giraffe may be less than the processing cost per kilogram of guinea fowl. But game animals differ substantially from many plant foods in that processing is such a small component of total cost that even a systematic difference by prey size may make a negligible difference in return rates.

References

Hart, J. A. 1978 From subsistence to market: a case study of the Mbuti net hunters. *Hum. Ecol.* **6**, 325–353.

Hill, K., Hawkes, K., Kaplan, H. & Hurtado, A. 1984 Seasonal variance in the diet of Ache hunter–gatherers in eastern Paraguay. *Hum. Ecol.* **12**, 101–135.
Ledger, H. P. 1968 Body composition as a basis for a comparative study of some east African mammals. *Symp. Zool. Soc. Lond.* **21**, 289–310.

R. J. Blumenschine, (*Department of Anthropology, Rutgers University, New Brunswick, New Jersey, U.S.A.*). Might the currency optimized in large mammal acquisition not be meat (i.e. protein) but rather energy from fat deposits around organs and in marrow bones? As a corollary, does the hunter have preferential access to fat deposits from his kill?

K. Hawkes. Fat may be the component of greatest interest. People talk about carcasses in terms of their fat content. Woodburn surmised that Hadza men had greater access to meat than women, and that the parts reserved for adult men gave them greater access to fat. Certain parts of large animals are reserved as 'men's meat' including, for example, the tongue, bellysheet, genitals, and foetuses. Speth (1990) follows Woodburn and uses an anecdote from Bunn *et al.* (1988) to illustrate discrimination against women in access to fat. Our data, covering a reasonable sample of cases, do not support this generalization. We do not have a quantitative answer to the question of preferential access to fat for the hunter, but men other than the hunter, as well as women and children, regularly consume both meat and fat at kill sites, including notably heads and the marrow in metapodials and long bones of several species (see O'Connell *et al.* (1991) for a tabulation).

One could translate our treatment, which uses kilograms of game, to strictly fat by choosing edible portion estimates for Hadza prey species and combining these with estimates of fat fractions by species and season. It could be that such substitutions throughout would change our results, but it seems unlikely. As long as the fat in big game is eaten by many while the fat in small game is not, men (and their families) would eat fat more often by claiming shares from any large prey taken but not specializing in taking it themselves.

References

Bunn, H. T., Bartram, L. & Kroll, E. 1988 Variability in bone assemblage formation from Hadza hunting, scavenging and carcass processing. *J. anthrop. Archaeol.* **7**, 412–457.
O'Connell, J., Hawkes, K. & Blurton Jones, N. 1991 Patterns in the distribution, site structure and assemblage composition of Hadza kill-butchering sites. *J. archaeol. Sci.* (In the press.)
Speth, J. D. 1990 Seasonality, resource stress, and food sharing in so-called 'egalitarian' foraging societies. *J. anthrop. Archaeol.* **9**, 148–188.

Comparative aspects of diet in Amazonian forest-dwellers

KATHARINE MILTON

Department of Anthropology, University of California, Berkeley, California 94720, U.S.A.

SUMMARY

Recent research shows that lowland forests of the Amazon Basin differ in numerous ways including features of climate and soils, faunal composition and forest structure, composition and phenology. Such differences strongly suggest that single-factor models used to explain features of human ecology in Amazonia may be too limited. A comparative study of the dietary ecology of four forest-living indigenous groups in Brazil (Arara, Parakana, Arawete, Mayoruna) revealed a number of differences. Primary crops, as well as animal types most utilized as prey, were found to differ markedly between groups. Although some differences can be accounted for by general environmental factors, no compelling single environmental factor can explain why any one group could not behave dietarily in ways more similar to another. Many of these intergroup dietary differences appear to represent a type of cultural character displacement that aids in distinguishing the members of one group from another. As all human groups, through the medium of culture, are actual or potential occupants of the same dietary niche, each group may distance itself from potential dietary rivals through cultural conventions. This behaviour may be justified, as the lack of overlap between forest-living groups in combination with generally intense intergroup hostility suggests that the biomass and distribution patterns of critical dietary resources in this environment may set limits to viable population size for particular areas.

1. INTRODUCTION

Single factor hypotheses related to diet have been invoked to explain numerous aspects of human ecology in tropical forests of the Amazon Basin, including population and settlement size, patterns of residence and nomadic and semi-nomadic behaviours (Meggers 1954, 1971; Carneiro 1960; Gross 1975). Limited dietary resources have been suggested to influence behavioural traits such as aggression, warfare and infanticide (Harris 1974; Chagnon & Hames 1979; Werner 1983). It has been hypothesized that tropical rain-forest environments are so nutrient-poor that human populations could not survive in them without access to crop foods (Headland 1987; Bailey *et al.* 1989). Tropical forest dwellers are often denigrated by terms such as 'marginal' peoples or 'refugees', the implication being that no one would choose to live in such an environment if other alternatives were possible (Lathrap 1968). All of this creates a largely negative and greatly oversimplified picture of tropical-forest peoples and their environment.

One common problem with these explanations has been the fact that most have been advanced without detailed knowledge of the potential of local dietary resources or actual food habits of the associated indigenous inhabitants (Hames & Vickers 1983). Often present-day environmental or dietary conditions are ascribed to the past without good historical or archaeological evidence. I suggest that all of the single-factor dietary models used to explain features of human ecology in forests of the Amazon Basin may have been pre-destined to failure by their attempts to provide a single generic explanation for what is, in effect, a vast and complex mosaic of different dietary possibilities calling for a variety of different solutions by human foragers (Dwyer 1986). These differences include: (i) wide regional and seasonal variation in rainfall patterns and hours of solar insolation; (ii) a range of soil conditions as diverse as in temperate zones (Sanchez & Buol 1977); (iii) a number of forest types with structure, composition and phenological production patterns which vary within, as well as between, geographical regions (Balee 1989; Gentry 1990; Bodmer 1990); and (iv) variation in the composition, distribution patterns and biomass of the associated faunal communities (Janzen 1974; Bodmer 1990). Given all of these differences, it seems logical to assume that the human inhabitants of such forests might face a wide range of different possibilities with respect to dietary potential.

2. THE MONKEY MODEL

My interest in the dietary ecology of neotropical forest dwellers was stimulated by earlier work on the diets of non-human primates in this same environment (Milton 1981, 1987). In the neotropics, non-human primates are confined to the forest canopy where the

Phil. Trans. R. Soc. Lond. B (1991) **334**, 253–263
Printed in Great Britain

largest biomass of digestible plant matter occurs. By specializing on different subsets of the available plant resources, generally supplemented by second trophic level foods, a large number of primate species are able to coexist sympatrically (Hershkovitz 1977; Terborgh 1983).

In non-human primates, strong intraspecific aggression between groups is the norm; as members of the same species, they occupy the same dietary niche, and intense dietary competition is expected to prevail. In contrast, interspecific aggression between sympatric groups of non-human primates is relatively rare because each monkey species has its own dietary niche.

In contrast to non-human primates, humans live on the forest floor where the biomass of edible plant matter is generally low. Similar to conspecific monkey groups, human groups in the Amazon Basin show a strong tendency to repel or discourage other human groups from utilizing their supplying area; this antipathy generally holds whether non-residents speak the same or a different language than residents. In ecological terms, it would appear that the establishment of one human group in a given area has a decidedly negative effect on the probability of establishment of another human group within the same area. Given all of the disturbances to indigenous populations in the Amazon Basin since contact, this spacing pattern may now be less obvious than in the past. However, the large number of different language groups known to have occupied forests of the Amazon Basin pre-contact (Nimuendaju 1987), in combination with the present-day locales of remaining groups (Lizarralde 1991), strongly suggests that in the past (and, in many areas, even today) Amazonian lowland forests were well saturated with indigenous groups, each of which strived to maintain exclusive use of its particular supplying area (Roosevelt 1980; Balee 1984).

The spacing behaviour of humans in forests of the Amazon Basin thus suggests that some features of human ecology or behaviour in this environment prohibit or minimize the potential for overlap. By analogy with non-human primates, dietary factors would appear to be implicated. To understand better what such factors might be, I here examine the respective forest environments and diets of some forest-living groups.

3. STUDY GROUPS

These forest-based groups have lived for many successive generations within the vast expanses of upland terra firme forest and should not be confused with other indigenous inhabitants of the Amazon Basin, such as riparian fisher–gardeners (e.g. the Tukanoans) or savannah–transition woodland peoples (e.g. the Kayapo). The inhabitants of these environments differ from one another in a large number of traits, including physical traits (Milton 1983), and typically reside in quite different ecological zones.

Although they were forest-dwellers, all the groups I worked with not only hunted and gathered but also practised slash-and-burn horticulture and are best regarded as hunter–gatherer–horticulturalists (HGH). Forest-based peoples should not be viewed as ahistorical, static isolates (see, for example, Schire 1984; Vansina 1990). Rather, all such groups have obviously been affected over centuries by contact both with other, generally hostile, indigenous groups and by outside influences, including trade goods, new cultivars and new diseases (Posey 1987; de Castro & de Andrada 1988; Roosevelt 1989). None the less, it is still possible to find little-acculturated groups who are long-term inhabitants of forests of the Amazon Basin, who still hunt almost exclusively with their traditional weapons, and whose food (possibly excluding salt) is still obtained through their own efforts and interactions with their forest environment. The four groups discussed in this paper conform to this description.

Three of the four study groups – the Arara, Parakana and Arawete – live in central Pará state, Brazil (figure 1). What is known of their history suggests considerable transitional movement throughout this region, such that in the past, one group may actually have lived in, or very near, an area now occupied by another. Actual hostilities are known to have taken place between at least two and possibly all three of these groups, in one case (between the Arawete and Parakana) as recently as 1983 (Arnaud 1983; de Castro 1988; de Castro & de Andrada 1988). The fourth group, the Mayoruna, lives to the west of the other three in the state of Amazonas (figure 1).

The Arara are Carib speakers presently settled north of the Iriri River (figure 1). They have lived in Pará for centuries, presumably having migrated south from the Brazil–Guiana region where a number of Carib-speaking groups still occur (de Castro & de Andrada 1988). Once estimated to consist of more than 300 individuals, the total group today consists of around 90 people. My work was done in Curambe, a small village on the Iriri River inhabited by 20 Arara, ten of whom were adults. During my study, the residents of this village were moved by the Indian Bureau (FUNAI) to a new site called Ikopty, closer to the junction of the Iriri and Xingu Rivers and the larger Arara village of Laranjal.

The Parakana, a group of approximately 350 individuals, live in three villages in the region between the Tocantins and Xingu Rivers. They speak a language of the Tupi–Guarani linguistic family. The village I worked with was located approximately 30 km east of the Xingu River on a small tributary called Bom Jardim (figure 1). This village is unusual in that until two years before my study its inhabitants had lived for 20 or more years as nomadic hunter–gatherers in the forest in an effort to escape attacks from hostile Gê-speakers and outsiders. In 1983, after being contacted by the Indian Bureau, 105 Parakana agreed to settle in the Bom Jardim area, and in 1984 they were joined by another 31 nomadic Parakana (Magalhães 1988).

The Arawete speak a language of the Tupi–Guarani linguistic family. This group used to live further east in the vicinity of the Bacaja River (de Castro 1988). Hostile attacks by the Xicrin–Kayapo apparently forced them to migrate west, displacing the Assurini in

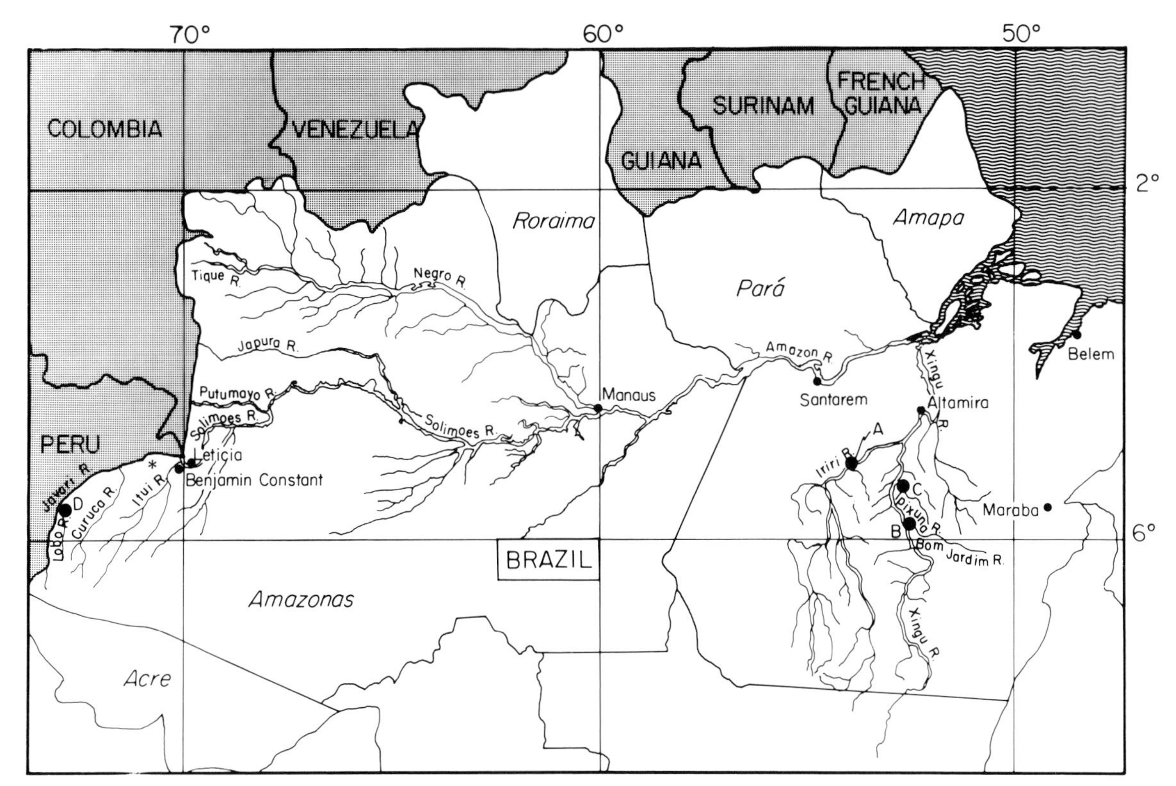

Figure 1. Study locales: A, Arara; B, Parakana; C, Arawete; D, Mayoruna. Note: * indicates tree-collection site for the Mayoruna area.

the process. Currently the entire group consists of some 110 individuals in a single village on the Ipixuna River, a small tributary east of the Xingu River (figure 1).

The Mayoruna, Panoan speakers, live east and west of the Javari River in Peru and Brazil. This group is noted for its extremely aggressive behaviour; its raids on settlers and river boats, and abduction of women, made the Javari River region unsafe for outsiders until the 1970s. The Mayoruna group I studied consisted of 117 individuals living in the state of Amazonas on the Lobo River, a small tributary of the Javari River (figure 1).

4. RESEARCH PROTOCOLS

(*a*) *General*

Three groups were visited on two occasions to compile data at different points in an annual cycle. Time constraints permitted only one visit to the Arawete. At each locale I collected environmental data, including daily rainfall and temperature range, soil samples, floral samples and phenological information. Data compiled on group members included height, mass and dental information. Activity budgets for adult men and women were recorded. Game and plant foods seen entering each village (or selected households) were identified and weighed. Fields were measured, crop types recorded and harvest mass obtained. Space constraints prohibit detailed discussion of all sampling protocols. Information on specific protocols relevant to data presented in this paper is given below.

(*b*) *Environmental factors*

Local climatic features were monitored by a maximum–minimum thermometer and standard rain gauge read at the same time each day. Mean annual rainfall and hours of solar insolation (see figure 2) for the two main study areas (central Pará and western Amazonas) were obtained from SUDAM, a Brazilian environmental agency. Soil collected at each site included samples from: (i) freshly burned fields; (ii) fields under cultivation for around 2 years; (iii) forested areas adjacent to fields; and (iv) undisturbed forest areas. Soil samples were analysed for several standard features by Agro Services International, Orange City, Florida. Forest composition was sampled by 0.25 or 0.5 ha† sample plots laid out in undisturbed terra firme forest in each study site. Within each plot, all trees greater than 10 cm d-b-h were measured and tagged. For three sites (Arara, Parakana and Mayoruna), professional tree collectors obtained botanical specimens from sample plots; most specimens were later identified to genus and species.

(*c*) *Diet*

As the collection of dietary data was a central focus of my study, I brought in my own food supplies so that local inhabitants would not have to secure food for me nor hide food for fear I would eat it. In each village I recorded the identity of all foods I saw and weighed all plant foods and game to which I could obtain access. I particularly tried to determine which crops and prey items were most frequently utilized in the daily diet.

† 1 ha = 10^4 m^2.

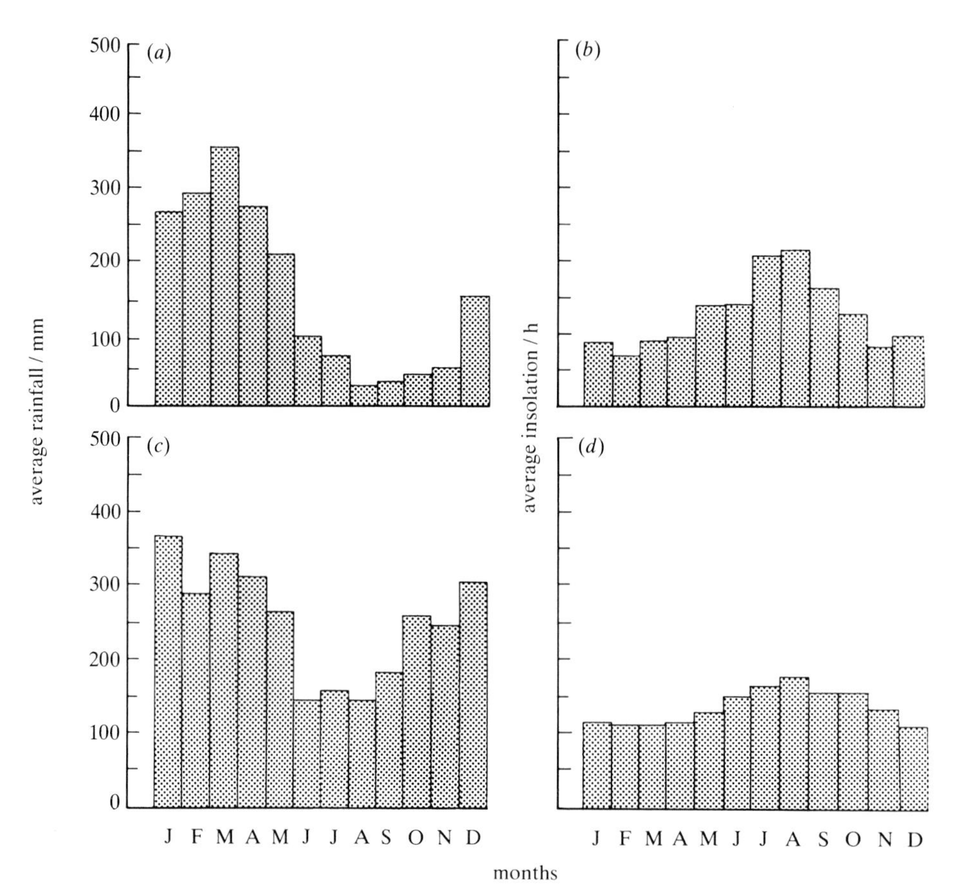

Figure 2. Averages over 20 years for monthly rainfall and solar insolation patterns for the two main study areas, central Pará state and western Amazonas state, Brazil. A and B: central Pará, station-Marabá; lat: 05°21′S; long: 49°09′W; altitude 102 m. C and D: western Amazonas, station-Benjamin Constant; lat: 04°23′S; long: 70°02′W; altitude 80 m. Figures redrawn after those presented in Atlas Climatológico da Amazônia Brasiliera, Ministerio do Interior, SUDAM/PHCA, Belém, Pará, 1984.

These I termed primary dietary resources, subdivided into two classes, primary carbohydrates and primary prey, defined for purposes of this paper as follows.

The primary carbohydrate for each group was defined as the plant food estimated to provide the highest percentage of calories in the daily diet on an annual basis. Because my stay in each village was substantially shorter than an annual cycle (table 1), I made this decision based on data from examination of fields and observations of plant food consumption. Indigenous informants and Indian Bureau employees also supplied information on seasonal dietary habits.

Primary prey for each group were defined as the animal types most frequently eaten. To determine primary prey I examined hunt return data from each group visit, and sorted prey into specific categories. I then totalled the number of prey items seen for that visit and calculated each prey category as a percentage of the total. Though the mass of many prey items was recorded, as well as the number of hunters and hours per hunt, discussion of these data is beyond the scope of this paper.

4. RESULTS

(*a*) *Variation in environmental factors*

Monthly data on rainfall and solar insolation patterns presented in figure 2 confirm regional and seasonal differences in these parameters for the two main study regions. Similar differences were noted for my on-site rainfall data. No general statements can be made with respect to soil analyses other than to note confirmation of the well appreciated fact that clearing and burning forest cover greatly enriches soil fertility for periods lasting longer than two years. These results were consistent regardless of locale. However, in the total data set, some in-site differences in soil parameters were as profound as between-site differences. Space does not permit discussion of these often complex results, which are presented in a separate publication (K. Milton, in preparation). Information on forest composition is presented in table 1. At the family level, forest composition did not differ notably between sites; there were, however, striking local differences in the number of genera and species per family, as well as the number of stems. For example, forest plots of the Mayoruna contained no arborescent palms; in contrast, palms were one of the best represented families in forest plots of the Arara and Parakana in Pará (table 1).

(*b*) *Variations in primary carbohydrates*

Important crops of the Arara included sweet potatoes (*Ipomoea batatas*), bitter manioc (*Manihot esculenta* Crantz), corn (*Zea mays*) and bananas (*Musa*

Table 1. *Forest composition by site*

tribe	number of stems per ha ⩾ 10 cm dbh	number of families	families with greatest number of genera		families with greatest number of stems		percentage of total stems in Palmae
Arara	467	37	Leguminosae	23	Leguminosae	114	13.1%
	composite of data		Moraceae	10	Bursuraceae	99	
	from four 0.25 ha		Crysobalanaceae	9	Palmae	61	
	plots from 3 areas		Annonaceae	8	Lecythidaceae	29	
	1–2 km apart		Bursuraceae	8	% of total = 64.9		
Parakana	397	41	Leguminosae	26	Palmae	63	15.9%
	composite of data		Moraceae	13	Leguminosae	45	
	from two 0.5 ha		Lauraceae	7	Moraceae	39	
	plots approximately		Flacortiaceae	7	Bursuraceae	28	
	1 km apart		Bursuraceae	7	Rutaceae	22	
			Meliaceae	7	Meliaceae	21	
					% of total = 54.9		
Arawete	408	n.d.	n.d.		Meliaceae	22[a]	3.9%[b]
	estimated from				Sterculiaceae	20[a]	
	data on one 0.25 ha				% of total (102		
	plot which				stems) = 41.1		
	contained 102 stems						
Mayoruna	582	42	Leguminosae	21	Lecythidaceae	74	0.0%
	composite of data		Crysobalanaceae	18	Crysobalanaceae	57	
	from two 0.5 ha		Sapotaceae	18	Myristicaceae	54	
	plots approximately		Myristicaceae	15	Leguminosae	49	
	1 km apart		Moraceae	11	Sapotaceae	41	
			Euphorbiaceae	11	Morasceae	39	
					Euphorbiaceae	38	
					% of total = 60.5		

[a] stems of these genera noted to be abundant in the one 0.25 ha plot surveyed.
[b] 3.9% Palmae in 0.25 ha plot.

spp.). Corn was a highly seasonal resource, whereas sweet potatoes, manioc and bananas were more or less continuously available. The primary carbohydrate of the Arara was a fermented beverage known colloquially as 'pik-tu'. It can be manufactured from almost any crop, but the Arara preferred sweet potatoes, presumably because they can be converted into a beverage far more rapidly than manioc or corn, both of which require extensive and time-consuming preparation.

While living as nomads, the Parakana made a type of gruel or bread from mesocarps of the babaçu palm nut (*Orbignya* sp.) and infructescences of the 'banana brava' plant (*Phenakospermum guianensis*). My phenological data show that ripe babaçu nuts are available in the forests of Pará throughout the year; informants stated that banana brava also fruits throughout the year.

The primary crop of the Parakana is bitter manioc (*Manihot esculenta* Crantz). The Parakana at Bom Jardim had planted manioc and consumed it during my study. Manioc roots are soaked for approximately three days in the river and peeled. The soft, water-soaked roots are squeezed into balls by hand which are then dried on a rack over a fire. The resulting material is crumbled through a sieve and used to prepare manioc bread or, occasionally, farinha (manioc cereal). Manoic and babaçu products comprised almost the total carbohydrate substrate for the Bom Jardim Parakana during both visits.

The Arawete were the most unusual group in that their primary carbohydrate was corn (*Zea mays*), which was consumed year-round rather than seasonally. De Castro (1988) states that the Arawete are the only Tupi group that cultivates corn for year-round consumption; they are the only forest-based group I have ever observed which has this custom. Each year, they clear large areas (one hectare or more) to make new corn fields; these are not necessarily placed on flood plains but can be found well inland. Once mature, corn is tied with vines in palm frond bundles, stacked on logs, covered and left in the fields. Every few days, groups of men travel to the corn stack, collect baskets of corn, each averaging approximately 22 kg, and bring them back to the village. Corn is removed from the cob, roasted on a large griddle, pounded to a powder in a wooden mortar and stored in tightly woven baskets.

The Arawete eat powdered corn throughout the day. A type of gruel is commonly prepared from corn and probably fresh corn is eaten roasted. There is no important secondary food crop for the Arawete: corn is the dietary keystone in terms of caloric intake. Balée (1989) has estimated that 82% of land under cultivation by the Arawete is devoted to corn, and this figure probably would be even higher if rice had not recently been introduced.

To quantify the importance of corn in the Arawete diet, I did some spot sampling of dietary items over a period of days, taking equal numbers of samples for each hour between 0800h and 1800h. Of 284 scans, corn powder accounted for 46% and corn gruel 2%.

Table 2. *Prey items recorded for each group (number and %[a] of total for that sample period)*

group		tortoise		bird		fish[b]		monkey		peccary		agouti		paca		armadillo		tapir		deer[d]		larvae		other	
		number	%	number	%	number	%	number	%	number	%	number	%	number	%	number	%	number	%	number	%	number	%	number	%
Arara	1985 (1)[c]	14	16	15	17	24	28	17	20	8	9	—	—	4	5	—	—	—	—	—	—	—	—	5	6
	1986 (2)	4	11	6	17	8	23	10	29	—	—	2	6	—	—	1	3	—	—	—	—	—	—	4	11
Parakana	1986 (3)	80	44	7	4	5	3	—	—	10	6	4	2	30	17	32	18	1	< 1	4	2	7	4	—	—
	1986 (4)	22	28	5	6	14	18	—	—	2	3	3	4	12	15	6	8	1	1	—	—	2	3	9	2
Arawete	1986 (5)	59	35	70	41	6	4	3	2	4	2	13	8	1	< 1	9	5	1	< 1	3	2	—	—	—	—
Mayoruna	1986 (6)	2	11	1	5	2	11	5	26	2	11	—	—	1	5	—	—	1	5	1	5	—	—	4	21
	1987 (7)	5	8	6	9	13	20	8	12	23	35	2	3	3	5	1	2	—	—	1	2	—	—	3	5

[a] All percentages rounded to the nearest whole number.
[b] Fish: each return with fish; 1 fish or many scored as 1; mass of fish returns generally were recorded.
[c] (1) 7 October–11 December; (2) 20 April–8 May; (3) 10 March–11 April; (4) 28 July–20 August; (5) 12 April–9 May; (6) 27 November–7 December; (7) 15 February–22 March.
[d] Generally killed for non-Indian consumption or believed to be a post-contact adaptation.

Rice (by chance being harvested) accounted for 27% of the scans. If my sample had not coincided with the rice harvest, I feel confident that corn products would have made up more than 75% of the total sample.

For each group, data were compiled on the size and condition of the residents' teeth. The Arawete had extremely poor teeth relative to other groups, who generally had excellent teeth. Individual Arawete as young as 19–24 years of age had molar teeth missing, as well as some premolars, and conditions only worsened with age. The poor condition of Arawete teeth may result from abrasive action of corn powder on dental enamel and the continuous decay of corn starch at the gum line.

The Arawete drank, and cooked corn only in fresh water from water holes which they dug in a clay-based substrate. They also ate clay from the bottom and sides of these water holes. Chemical attributes of this water may enhance nutrients found in corn, or aid in detoxification of adverse chemicals in the Arawete diet (see, for example, Johns 1990). A non-dietary possibility is that, by drinking only ground-filtered water, the Arawete may avoid water-transmitted parasites.

The Mayoruna generally place their large fields on hills rather than flat ground. They were the only group observed to weed fields. Their primary crop is sweet manioc (*Manihot esculenta* Crantz), which is boiled and eaten or processed into farinha. This farinha is yellow in colour, in contrast with the pale cream colour of farinha manufactured from bitter manioc. The second most important crop of the Mayoruna is bananas, particularly plantains. In summary, each group utilized a different primary crop (table 3).

(c) *Variation in primary prey*

Primary prey also showed wide variation between groups. The Arara consumed the widest range of prey species, including sting rays and electric fish; the intestinal tract of a wide variety of mammals was routinely consumed. Pinto (1989) reports that vultures, house rats and hawks are eaten by the Arara when other meat is not available. The single most important prey item of the Arara was monkeys, particularly capuchin monkeys (*Cebus apella*) (table 2). In samples, monkeys accounted for 20–29% of the total prey brought back to the village. This percentage is a minimum estimate as I was not able to quantify individually 38 kg of smoked game, largely monkeys, brought back to the village for a festival. The Arara maintain large numbers of monkeys in the village as pets, obtained when their mothers are killed for food. The study group also included a notable amount of fish in the diet (table 2). The dietary importance of fish appears to be recent, resulting from settlement of the Arara near the Iriri River by the Indian Bureau who also provided hooks and lines. At the larger Arara village, Laranjal, fishing is regarded as an activity of little value (Pinto 1989).

In striking contrast to the Arara, the Parakana are specialists on terrestrial game, particularly land tortoises, tapir and wild pigs, but also armadillo and paca (table 2). They hunt with dogs, which is effective for

Table 3. *Summary of dietary differences*

group	primary carbohydrate	primary prey	avoided prey
Arara	sweet potato	monkeys	deer tapir disliked
Parakana	bitter manioc	tortoises	deer monkeys cracids and macaws
Arawete	corn	large birds	deer tapir disliked
Mayoruna	sweet manioc	peccaries	Approximately 12–16 prey types including snakes, felids, anteaters, coatis, various monkey and bird species, squirrels, etc. Deer were probably avoided pre contact

terrestrial game. The emphasis on land tortoises in the diet was further confirmed by examination of an indigenous rubbish dump at Bom Jardim. I counted 313 largely intact tortoise shells as well as innumerable shell fragments.

However, when family groups left the village to go on hunting treks in the forest (as they frequently did), hunters stated very emphatically that treks were undertaken with the specific intent of hunting tapir. Remains of smoked game brought back to the village confirm that the Parakana generally were highly successful in this pursuit. Tapir meat was the most common, and frequently the only, food brought back to the village after treks, often in amounts weighing 30–45 kg.

Animal species well represented in the Parakana diet tended to have notable fat stores, particularly at certain times of year (Speth & Spielmann 1983). For example, in March 1986, I estimated that tapirs had a layer of fat greater than 2 cm thick beneath the skin. I observed the Parakana eating chunks of tapir and paca fat, and they are reported to take fish fat, mix it with farinha (manioc cereal) and eat this. Informants stated that the Parakana had been observed to gorge on tapir fat and meat to the point of illness.

The Parakana also eat insects, particularly palm larvae which are a rich source of fat. Analyses show fat contents (dry mass) from 35% to 69% for various types of palm larvae, which also tend to be high in protein (range 22–39% dry mass (K. Milton, unpublished data)). When living as hunter–gatherers in the forest, animal fat may have been particularly important with respect to calories for the Parakana, as wild plant foods may at times have been scant. Clastres (1972) noted that the non-horticultural Guayaki actively cultivated palm larvae at sites throughout the forest, such that this energy-rich dietary resource was available to them throughout the year.

The Arawete were the only group I worked with who kept no dogs, nor did they like or desire them. Perhaps related to this, much of their hunting activity resulted in large game birds such as moutons (*Crax fasciolata*, *Mitu mitu*), jacoos (*Penelope jacquacu*), macaws (*Ara macao*, *A. chloroptera*, *A. ararauna*, *Anodorhynchus hyacinthius*) and toucans (*Ramphastos tucanus*, *R. vitellinus*) (table 2). The Arawete village was densely populated by pet macaws of the above-mentioned species, as well as many smaller hook-bills.

As shown in table 2, the Arawete also ate a variety of other prey items including numerous land tortoises as well as agoutis and monkeys. De Castro (1986) regards tortoises, particularly tortoise liver, as the most preferred animal food of the Arawete. He also notes that the Arawete possess some 45 classifications for types of honey.

Although the Mayoruna ate a range of prey species, hunting returns show a focus on wild pigs (*Tayassu* spp.) and certain of the larger monkey species (table 2). Wild pigs were taken more frequently and consistently by the Mayoruna than by any of the other three groups; of the 25 pigs killed, 19 were collared peccaries (*T. tajaca*) and six were white-lipped peccaries (*T. pecari*). Although only one tapir was killed during my study, the Mayoruna stated that tapir was an important and preferred prey species.

The Mayoruna were the only group I saw eating sloths (*Choloepus* sp.); they stated that sloths were one of their most preferred foods. Sloths are captured by climbing their tree, lassoing them with a noose made of vines, pulling them free and then clubbing them to death on the ground. Other game is secured with bow and arrow. Fish were also important in the Mayoruna diet (table 2). The Mayoruna had been at Lobo for over nine years and stated emphatically that 'they would never run out of game' in this locale though eventually 'they might have to travel longer distances to secure it'.

(*d*) *Variation in avoided prey*

The three Pará groups had few prey restrictions (see table 3). None of them apparently ate deer before European contact. Deer were stated to be avoided because they lacked fat reserves and also 'were spirits'. Before European contact the Arara also avoided tapir as food; it was said to be undesirable, 'strong' meat. Ross (1978) notes that the Achuara consider deer and tapir to be reincarnated spirits; eating tapir is also believed to cause a skin rash. Before European contact the Parakana did not eat any monkeys or deer (most still do not), and largely avoided eating larger birds such as macaws (*Ara* spp.) and moutons (*Crax* spp., *Mitu* spp.). The Arawete dislike tapir meat, although apparently they will eat it.

In contrast, the Mayoruna had an extensive number of avoided prey. This list consisted of 12–16 different prey types, including snakes, porcupines, squirrels, felids, parrots, macaws, anteaters (two species), tyras, capybaras, coatis and various monkey species including capuchin monkeys and uacaris; adult Mayoruna do not eat howler monkeys although children do. The Mayoruna also stated they would not eat fish without scales or insects other than those found on the human body. Some Mayoruna were observed to eat deer but the antiquity of this practice is unknown.

6. DISCUSSION

(*a*) *Factors related to dietary differences*

Certain broad and obvious environmental differences between regions appear to affect the dietary pattern of their indigenous inhabitants. The Mayoruna, for example, live in an area lying outside the ancient and heavily weathered Guianan and Brazilian shields (Lathrap 1970). Climatic data confirm high and relatively uniform rainfall, and forest samples confirm a high and relatively even diversity of tree types per unit area. High primary productivity and less dominance by particular plant groups should result in a diverse and high faunal biomass which, in turn, should be reflected in the dietary pattern of the Mayoruna.

Among the Mayoruna, prey choice was more specialized than in other groups as hunters generally were able to secure large and highly preferred prey species. Conversely, in Pará the return on large game was low and less even than the return on small game (table 2). Prey selection by Mayoruna appeared similar to that of the Waorani, who live in a productive lowland forest region in eastern Ecuador. Large monkeys, peccaries and large birds make up the highest percentage (seven species compose > 53 %) of Waorani kills (Yost & Kelley 1983). Further, in striking contrast to the three Pará groups, the Mayoruna had an extensive list of avoided prey. Although sweet manioc was their most important crop for calories, the Mayoruna also routinely consumed large quantities of bananas, so much so that, in effect, they came close to having two primary carbohydrates. In contrast, the Pará groups were more narrowly focused on a single carbohydrate crop. Overall, the Mayoruna showed a narrower prey focus and a broader crop focus than other groups considered in this study, a pattern that appears related to their generally more productive environment.

Yet beyond such very general and obvious environmental differences, none of the environmental factors I examined appeared sufficiently distinct to explain most of the dietary differences observed (although, certainly, future research on this question may provide new data to alter this view). *A priori*, based on present data, in terms of the physical environment, it seems probable that any one group could successfully cultivate the primary carbohydrate crop of another. In contrast to this view, Balée (1989) has suggested that intensive corn cultivation by the Arawete is made possible by the presence of unusually fertile *terra preta* soils in their geographical region. In my view the success of the Arawete as corn cultivators does not stem from the patchy presence of *terra preta* soils in this region, but rather from the Arawete habit of clearing large new fields for corn cultivation each year, a labour-intensive endeavour, particularly before the acquisition of steel tools.

It is also difficult to determine why each group focused on (or avoided) somewhat different prey types (for discussion of this loaded topic see Sahlins 1976; Ross 1978; Harris 1987; Vayda 1987). For example, I can find no environmental reason why the Parakana do not eat monkeys. They have lived in Pará for centuries, a state in which monkeys are abundant. It is difficult to argue that the heavy, broad arrows of the Parakana preclude or make difficult the hunting of monkeys, as the arrows of the Arawete seemed almost identical to those of the Parakana, yet the Arawete focus much of their hunting on birds and also eat some monkeys. Deer and tortoises occur in the supplying areas of all four groups and yet deer were avoided whereas tortoises were eaten. The Parakana and Mayoruna strongly desire tapir meat, whereas the Arawete and Arara find tapir meat barely edible. The Arawete specialization on large birds could indicate that birds are unusually abundant in their area; it could also indicate that the Arawete preferentially hunt birds or that there are few other prey types. Having corn as a year-round staple may reduce their need for protein from game. A better understanding of environmental effects on patterns of prey selection could be obtained through censuses to determine the relative densities of different prey types in each group's supplying area.

The different prey choices and dietary habits of each group appear in large part to reflect features of their origins, history and experience in combination with present-day ecological constraints. Although recognizing that explanations of foodways are generally neither simple nor clear-cut (see, for example, Sahlins 1976; Vayda 1987; Harris 1987), I can find no compelling environmental reason why any one group could not behave dietarily in ways more similar to another if it so desired. These different dietary practices appear to reflect, as do the distinctive facial perforations and body decorations of each group, a type of cultural character displacement in which the members of group A seek to differentiate themselves from members of group B or C by means that are distinctive but do not pose any actual economic disadvantage. As Rozin (1976, p. 65) has noted, dietary customs help to define cultural and social categories and draw distinctions within and between groups. Indeed the Indian name possessed by these groups typically translates into the 'true' or 'real' people, with outsiders described as not quite human (Birket-Smith 1965). Similarly, Bahuchet *et al.* (1990), in discussing the different agricultural strategies found among four non-Pygmy groups in the Lobaye forest, comment 'by investing with cultural values one or more plant species within their food system... each of these contiguous groups affirms its own ethnic identity' (Bahuchet *et al.* 1990, p. 33; see also Dwyer 1986; Johnson & Baksh 1987).

(*b*) *Returning to the monkey model*

Data presented above show that forest-based groups in the Amazon Basin have many dietary differences. Despite these differences, human groups in these forests do not live sympatrically; rather, allopatry is the decided norm. Sympatric monkey species often show notable differences in details of their dental and digestive morphology and physiology (Milton 1981). These adaptations appear to function such that each monkey species can dominate a particular subset of the available dietary resources (Milton 1981; Terborgh 1983). Data suggest that, not infrequently, these interspecific differences are sufficiently pronounced such that successful invasion of one monkey species into the dietary niche of another may not be possible (Milton 1981).

This is not the case for human groups. Basically, all humans have the same digestive morphology and physiology, as well as the same need for high-quality dietary resources (Milton 1987). Humans in tropical forests meet their dietary needs with a wide range of food items, but food choices tend to follow a common theme, that of using plant foods to meet most energy demands and animal foods to satisfy most protein demands. Humans may form into discrete social units with particular dietary customs and term themselves 'the people', they may decorate their faces and bodies with specific group symbols but, unlike members of different monkey species, no human group is immune to dietary invasion by members of any other group and the immediate adaptation of any and all aspects of its subsistence behaviour that seem desirable to the invaders. Through the adaptive mechanism of culture, all human groups are actual or potential occupants of the same dietary niche.

However, this fact alone does not appear sufficient to explain the allopatry and general xenophobia characteristic of these forest-based groups. Rather, such behaviour suggests that some essential foods in the tropical forest may be distributed in patterns that set limits to viable population size (greater or lesser, depending on features of forest productivity at particular sites). An increase in human biomass in a given supplying area might therefore overtax local dietary resources or make their acquisition prohibitively costly (central place foraging, for example, sets an upper limit on viable day range).

It could be argued that these different dietary habits could serve as mechanisms to facilitate spatial overlap by different forest-based human groups, and that in the past more intergroup overlap might have prevailed. There seems little evidence, either historical or recent, to support this assumption. Not only do present-day groups show a high degree of intolerance for individuals of other linguistic backgrounds, but there is frequently considerable hostility between groups speaking the same language (see, for example, Arnaud 1983; Balée 1984; K. Milton, personal observation). Roosevelt (1980, p. 35) has remarked specifically on the possibility of pre- and post-contact local and regional population pressures on subsistence resources. As she notes, the frequent association of female infanticide, polygyny and warfare suggests the existence of a self-reinforcing system of population control. Mechanisms of population control suggest that particular supplying areas may at times approach the limits of their carrying capacity (Roosevelt 1980).

Before horticulture, forest-living groups presumably relied on wild plant foods for much of their caloric intake (see, for example, Milton 1984). Advantages for human populations utilizing plant foods, rather than meat, as an energetic substrate are discussed in Milton 1984 (pp. 19–20). In a remarkable recent paper, Dwyer & Minnegal (1990) suggest that the present-day hunting patterns of the Kubo in lowland rain forests of New Guinea relate not to present-day horticultural practices but rather to distribution patterns of their pre-horticultural carbohydrate staple, sago palms (*Metroxylon* sp.). Similarly, the Guaja Indians of Brazil have lived for many generations as nomadic hunter–gatherers (FUNAI archives, unpublished report). An informant who lived with one Guaja group for 14 months after first contact stated that their travel patterns revolved around stands of babaçu palms, the mesocarps and kernels of which provided the bulk of their intake with respect to calories. Spacing patterns of human groups in the tropical forest might, therefore, have evolved initially to protect access to limited and highly desirable caloric supplies of wild plant foods. Once horticultural practices were adopted, protection of wild carbohydrate foods should not be critical. However, with the adaptation of horticultural practices, population size of hunter–gatherer groups presumably increased, as a result of improved and more reliable caloric returns (see, for example, Milton 1984). Pressures on prey resources may then have intensified to provide protein and animal fat for this greater human biomass. Thus, unless these present-day spacing patterns and avoidance behaviours are historical artifacts, it seems likely that the efficient utilization of patchily distributed game resources may be involved. The addition of extra-group hunters into a given supplying area could reduce the resident group's hunting returns, either through disruption of traditional patterns of in-group land use (which could lower prey capture rates or increase temporal or energetic expenditures involved in hunting) or by actually lowering the amount of prey available to in-group hunters. Greater knowledge of relative prey abundances and renewal rates as well as details of hunter movements over their total supplying area for one or more annual cycles would help to test these possibilities. Local group size, prey sharing networks and day range are critical factors that must be kept in balance with the carrying capacity of each supplying area (Dwyer & Minnegal 1990).

All of the various considerations discussed above make it clear that single factor hypotheses are unlikely to suffice in explaining the dietary and other ecological practices of Amazonian forest dwellers. Rather, in the future, considerably more attention should be paid to historical factors (such as linguistic affiliations, demographic shifts, migration patterns and differential access to new crops), local environmental conditions (particularly detailed faunal surveys) and general patterns of inter- and intra-group cultural dynamics.

This research was supported by grants from the National Geographic Society, the John Simon Guggenheim Memorial Foundation and the Miller Foundation for Basic Research in Science. I thank Julio Cezar Melatti, who served as my Brazilian sponsor for this study. Discussions with A. Wiley greatly helped to clarify many ideas in this paper, as did comments on the manuscript by P. Dwyer and A. Whiten.

7. REFERENCES

Arnaud, E. 1983 Mundanças entre grupos indígenas tupí da região do Tocanins-Xingu (Bacia Amazônica). *Bolm Mus. para. 'Emilio Goeldi', Nova Serie, (Anthropologia)* **84**, 1–50.

Bahuchet, S., Hladik, C. M., Hladik, A. & Dounias, E. 1990 Agricultural strategies as complementary activities to hunting and fishing. In *Food and nutrition in the African rain forest* (ed. C. M. Hladik, S. Bahuchet & I. de Garine), pp. 31–35. Paris: UNESCO/MAB.

Bailey, R. C., Head, G., Jenike, M., Owen, B., Rechtman, R. & Zechenter, E. 1989 Hunting and gathering in tropical rain forest; it is possible? *Am. Anthropol.* **91**, 59–82.

Balée, W. 1984 The ecology of ancient Tupi warfare. In *Warfare, culture and environment* (ed. R. B. Ferguson), pp. 241–265. New York: Academic Press.

Balée, W. 1989 The culture of Amazonian forests. *Adv. econom. Bot.* **7**, 1–21.

Birket-Smith, K. 1965 *The paths of culture.* Madison: University of Wisconsin Press.

Bodmer, R. E. 1990 Responses of ungulates to seasonal inundations in Amazon flood plain. *J. trop. Ecol.* **6**, 191–201.

Carneiro, R. L. 1960 Slash and burn agriculture: a closer look at its implications for settlement patterns. In *Man and cultures: selected papers of the Fifth International Congress of Anthropological and Ethnological Sciences* (ed. A. Wallace), pp. 229–234. Philadelphia: University of Pennsylvania Press.

Chagnon, N. A. & Hames, R. B. 1979 Protein deficiency and tribal warfare in Amazonia: new data. *Science, Wash.* **203**, 910–913.

Clastres, P. 1972 The Guayaki. In *Hunters and gatherers today* (ed. M. B. Bicchieri), pp. 138–174. New York: Holt, Rinehart and Winston.

de Castro, E. V. 1986 Araweté: Os Deuses Canibais. Rio de Janeiro: Jorge Zahar Editor Ltda.

de Castro, E. V. 1988 Os Araweté. In *As Hidrelétricas do Xingu e os Povos Indígenas* (ed. L. A. O. Santos & L. M. M. de Andrada), pp. 185–190. São Paulo: Comissão Pro-Índio de São Paulo, Zerodois Servicos Editoriais S/C Ltda.

de Castro, E. V. & de Andrada, L. M. M. 1988 Os povos indígenas do medio Xingu. In *As Hidrelétricas do Xingu e os Povos Indígenas* (ed. L. A. O. Santos & L. M. M. de Andrada), pp. 135–146. Sao Paulo: Comissão Pró-Índio de São Paulo, Zerodois Servicos Editoriais S/C Ltda.

Dwyer, P. 1986 Living with rainforest: the human dimension. In *Community ecology: pattern and process* (ed. J. Kikkawa & D. J. Anderson), pp. 342–367. Oxford: Blackwell Scientific Publications.

Dwyer, P. & Minnegal, M. 1990 Hunting in a lowland, tropical rain forest: toward a model of non-agricultural subsistence. Paper presented in the symposium 'Tropical forest ecology: the changing human niche and deforestation'. Annual meeting, *Am. Anthropol. Assn*, New Orleans 1990.

Gentry, A. H. 1990 Floristic similarities and differences between southern Central America and upper and central Amazonia. In *Four neotropical rainforests* (ed. A. H. Gentry), pp. 141–157. New Haven: Yale University Press.

Gross, D. R. 1975 Protein capture and cultural development in the Amazon Basin. *Am. Anthropol.* **77**, 526–549.

Hames, R. B. & Vickers, W. T. 1983 Introduction. In *Adaptive responses of native Amazonians* (ed. R. B. Hames & W. T. Vickers), pp. 1–28. New York: Academic Press.

Harris, M. 1987 Comment on Vayda's review of Good to Eat: riddles of food. *Hum. Ecol.* **15**, 511–518.

Headland, T. N. 1987 The wild yam question: how well could independent hunter–gatherers live in a tropical rain forest ecosystem? *Hum. Ecol.* **15**, 463–491.

Hershkovitz, P. 1977 *Living New World monkeys (Platyrrhini)*, vol. 1. University of Chicago Press.

Janzen, D. G. 1974 Tropical blackwater rivers, animals and mast fruiting by the Dipterocarpaceae. *Biotropica* **6**, 69–103.

Johns, T. 1990 *With bitter herbs they shall eat.* Tucson: University of Arizona Press.

Johnson, A. & Baksh 1987 Ecological and structural influences on the proportions of wild foods in the diets of two Machiguenga communities. In *Food and evolution* (ed. M. Harris & E. B. Ross). Philadelphia: Temple University Press.

Lathrap, D. W. 1968 The 'hunting' economics of the tropical forest zone of South America: an attempt at historical perspective. In *Man the hunter* (ed. R. B. Lee & I. de Vore), pp. 23–29. Chicago: Aldine.

Lathrap, D. W. 1970 *The Upper Amazon.* New York: Praeger Publications.

Lizarralde, M. 1991 *Grupos etno-linguistico autóctones de America del Sur: indice y mapa* (Suppl. 4). Caracas: Anthropologia.

Magalhães, A. C. 1988 O povo indígena Parakanã. In *As Hidrelétricas do Xingu e os Povos Indígenas* (ed. L. A. O. Santos & L. M. M. de Andrada), pp. 185–190. São Paulo: Comissão Pró-Índio de São Paulo, Zerodois Serviços Editoriais S/C Ltda.

Meggers, B. J. 1954 Environmental limitation on the development of culture. *Am. Anthropol.* **56**, 801–824.

Meggers, B. J. 1971 *Amazonia: man and culture in a counterfeit paradise.* Chicago: Aldine-Atherton.

Milton, K. 1981 Food choice and digestive strategies of two sympatric primate species. *Am. Nat.* **117**, 476–495.

Milton, K. 1983 Morphometric features as tribal predictors in Northwestern Amazonia. *Ann. hum. Biol.* **10**, 435–440.

Milton, K. 1984 Protein and carbohydrate resources of the Maku Indians of northwestern Amazonia. *Am. Anthropol.* **86**, 7–27.

Milton, K. 1987 Primate diets and gut morphology: implications for human evolution. In *Food and evolution: toward a theory of human food habits* (ed. M. Harris & E. B. Ross), pp. 93–116. Philadelphia: Temple University Press.

Nimuendaju, C. 1987 *Mapa etno-historico de Curt Nimuendaju, 1944.* Rio de Janeiro: Fundação Instituto Brasileiro de Geografia e Estatistica em colaboração com a Fundação Nacional Pro-Memoria.

Pinto, M. T. 1989 Os Arara: tempo, espaco e relações sociais em um povo Karibe. Unpublished Master's thesis in Anthropology, Federal University, Rio de Janeiro.

Posey, D. A. 1987 Contact before contact: typology of post-Colombian interaction with northern Kayapo of the Amazon Basin. Bol. Mus. Par. E. Goeldi, Ser. Anthrop. **3**, 135–154.

Roosevelt, A. C. 1980 *Parmana: prehistoric manioc and maize subsistence along the Amazon and Orinoco*. New York: Academic Press.
Roosevelt, A. C. 1989 Resource management in Amazonia before the conquest: beyond ethnographic projection. *Adv. econom. Bot.* **7**, 30–62.
Ross, E. B. 1978 Food taboos, diet and hunting strategy: the adaptation to animals in Amazonian cultural ecology. *Curr. Anthropol.* **19**, 1–36.
Rozin, P. 1976 The selection of foods by rats, humans and other animals. In *Advances in the study of behavior* (ed. J. S. Rosenblatt, R. A. Hinde, E. Shaw & C. Beer), pp. 21–76. New York: Academic Press.
Sahlins, M. 1976 *Culture and practical reason*. University of Chicago Press.
Sanchez, P. A. & Buol, S. W. 1977 Soils of the tropics and the world food crisis. In *Agricultural mechanization in Asia, Winter*, pp. 37–44. Amer. Assn. Advancement of Science.
Schire, C. 1984 *Past and present in hunter gatherer studies*. New York: Academic Press.
Speth, J. & Spielmann, K. 1983 Energy source, protein metabolism and hunter–gatherer subsistence strategies. *J. Am. Archaeol.* **2**, 1–31.
SUDAM 1984 *Atlas Climatológico da Amazônia Brasiliera*. Ministerio do Interior: Superintendencia do Desenvolvimento da Amazônia (SUDAM)/Projeto de Hidrologia e Climatológia da Amazônia, CDU 551.58(811) (084.4). Belém, Pará.
Terborgh, J. 1983 *Five New World primates*. Princeton University Press.
Vansina, J. 1990 *Paths in the rainforest*. Madison: University of Wisconsin Press.
Vayda, A. P. 1987 Explaining what people eat: a review article. *Hum. Ecol.* **15**, 493–510.
Werner, D. 1983 Why do the Mekranoti trek? In *Adaptive responses of native Amazonians* (ed. R. B. Hames & W. T. Vickers), pp. 225–238. New York: Academic Press.
Yost, J. A. & Kelley, P. M. 1983 Shotguns, blowguns and spears: the analysis of technological efficiency. In *Adaptive responses of native Amazonians* (ed. R. B. Hames & W. T. Vickers), pp. 189–224. New York: Academic Press.

Discussion

C. D. Knight (*Polytechnic of East London, U.K.*). Could the author speculate on the reasons for the widespread avoidance of deer meat? The ingroup–outgroup logic would not seem applicable in this case, as all four of the cultural groups – in addition to many others – evidently avoid eating deer.

K. Milton. This is a question that many anthropologists have tried to answer. In the tribes I worked with, the answer given for not eating deer meat was that deer were spirits and that deer lacked fat. The fact that deer meat is so widely avoided by so many different indigenous groups suggests that it may be an ancient avoidance. A discussion I had with Dr P. Dwyer has suggested a possible explanation for the widespread avoidance of deer meat. If we assume, as I believe, that indigenous groups in Amazonia were living in these forests well before the arrival of horticulture, there must have been periods each year when calorie-rich wild foods were relatively scarce. At such times, given the limitations of human physiology in terms of catabolizing amino acids for energy (Speth, this symposium), humans should have been particularly interested in seeking out foods containing either fats or carbohydrates to provide the calories they required each day. During these periods, lean meat offering little other than amino acids may have been avoided; gradually such seasonal avoidance may have developed into a general avoidance of deer as food. Bear in mind, however, that many of these food avoidances can be more apparent than real. In times of protein shortage or with moderate access to energy-rich wild foods, there may in fact be some means of relaxing this avoidance such that some or all members of the group can consume deer meat.

I. Crowe (*23 Lockhart Close, Dunstable, Bedfordshire, U.K.*). Given the primary carbohydrates mentioned were all species introduced into the forest (mainly from elsewhere in the New World), how reliable were the plant resources previously exploited, and were there any attempts to encourage propagations?

K. Milton. Indigenous peoples routinely disperse seeds of some forest species, for example *Inga* species and species of Palmae, in areas where they settle, but there is no way of knowing whether this practice is generally deliberate of accidental. Balée, for example, has suggested that at least 11.9% of the terra firme forests of the Brazilian Amazon are anthropogenic (Balée 1989). In terms of wild carbohydrate foods, I did a survey of the availability of ripe palm nuts of the babaçu palm, *Orbignya* sp. at various locales in Pará, and found these nuts to be available throughout the year. Seeds of *Phenakospermum guianensis*, the banana brava plant, are stated by indigenous informants to be available throughout the year, but I did not do any phenological surveys of this species. I would predict that certain other palm species, *Ficus* species etc., likewise produce fruit for most or all of an annual cycle. At least three edible wild roots, two small, in clusters and more like potatoes (identified as members of the Marantaceae and Araceae; *Dracontium* cf. *longipes* Engl.) and one very large root the size of a basketball, were shown to me by the Mayoruna. The large root in particular was stated to have been eaten before horticulture and still to be eaten occasionally. I would imagine that these are in the forest throughout the year, but their relative abundances per unit area or renewal rates are not known.

Protein selection and avoidance strategies of contemporary and ancestral foragers: unresolved issues

JOHN D. SPETH

Museum of Anthropology, University of Michigan, Ann Arbor, Michigan 48109, U.S.A.

SUMMARY

During seasonal or inter-annual periods of food shortage and restricted total calorie intake, ethnographically and ethnohistorically documented human foragers, when possible, under-utilize foods that are high in protein, such as lean meat, in favour of foods with higher lipid or carbohydrate content. Nutritional studies suggest that one reason for this behaviour stems from the fact that pregnant women, particularly at times when their total calorie intake is marginal, may be constrained in the amount of energy they can safely derive from protein sources to levels below about 25% of total calories. Protein intakes above this threshold may affect pregnancy outcome through decreased mass at birth and increased perinatal morbidity and mortality. This paper briefly outlines the evidence for the existence of an upper safe limit to total protein intake in pregnancy, and then discusses several facets of the issue that remain poorly understood. The paper ends by raising two basic questions directed especially toward specialists in primate and human nutrition: is this protein threshold real and demographically significant in modern human foraging populations? If so, does an analogous threshold affect pregnant female chimpanzees? If the answer to both of these questions is yes, we can then begin to explore systematically the consequences such a threshold might have for the diet and behaviour of early hominids.

1. INTRODUCTION

The origins of our earliest hominid ancestors, and the factors that led to that origin, have attracted the attention of scholars since the time of Darwin. The diet of these early bipeds has always been of prime interest, and quite understandably attention has most often focused on the role of hunting and meat eating (Gordon 1987; Hill 1982). The reasons for this focus are obvious. Animals are large packages of high-quality protein as well as calories. They are also among the most obvious high-quality food sources in the open semi-arid savannas of sub-Saharan Africa, where our earliest ancestors are believed to have first arisen (Klein 1989). In addition, consistent with expectations drawn from optimal foraging theory, animals are highly ranked among the food choices of modern hunters and gatherers, peoples who provide the inspiration for many of our current models about hominid origins and evolution (e.g. Binford 1978; Bunn *et al.* 1988; O'Connell *et al.* 1988; Winterhalder & Smith 1982). Moreover, even our closest living primate relatives, the chimpanzees, are now known to hunt and consume meat on a fairly regular basis, a strong argument that the common ancestors of both chimpanzees and humans probably also did so (Teleki 1981; Wrangham 1977). Finally, stone tools and fossil animal bones – the latter commonly displaying distinctive cut-marks produced when a carcass is dismembered and stripped of edible flesh with a sharp-edged stone flake – are found together on many Plio-Pleistocene archaeological sites, convincing proof that by at least 2.0 to 2.5 Ma before present (BP) these early hominids did in fact eat meat (Bunn 1986; Isaac & Crader 1981). In contrast, plant remains are absent or exceedingly rare on these ancient sites and their role in early hominid diet, therefore, can only be guessed on the basis of their known importance in contemporary forager diets, as well as their potential availability in Plio-Pleistocene environments (for example, see Peters *et al.* (1984); Sept (1984)). Thus few today doubt that early hominids ate meat, and most would agree that they probably consumed far more meat than did their primate forebears. Instead, most studies nowadays focus primarily on how that meat was procured: that is, whether early hominids actively hunted animals, particularly large-bodied prey, or scavenged carcasses that had already been partly consumed and abandoned by other predators (see, for example, Binford (1981); Blumenschine (1987, this symposium); Bunn (1986); Potts (1982); Shipman (1983)).

I fully concur with the view that meat was a regular and important component of early hominid diet. For this the archaeological and taphonomic evidence is compelling. Instead, what concerns me is how much of their energy, on average, was actually obtained from meat. And was the amount of meat in early hominid diet constrained just by their lack of adequate technological and organizational means for procuring large and often dangerous animals, or were there also intrinsic nutritional factors that, at least seasonally, may have limited the extent to which early hominids could rely on meat for calories, independent of the quantity that was potentially available on the landscape? Although there seems little doubt that technological and organizational constraints were important, there is growing evidence to suggest that nutritional constraints may also have been significant. As these nutritional arguments have been presented in detail elsewhere (Speth 1983, 1987, 1989, 1990; Speth & Spielmann 1983), I will only briefly outline them here, and use the remainder of this paper as an opportunity to identify specific critical facets of this 'excess protein' argument that remain poorly understood and that require for their solution further input from specialists in the fields of human and primate nutrition.

2. THE 'EXCESS PROTEIN' ARGUMENT

Whereas meat can obviously provide calories as well as protein, there is an upper limit to the total amount of protein (plant and animal combined) that one can safely consume on a regular basis. This limit – best expressed as the total number of grams of protein per unit of lean body mass that the body can safely handle – is about 300 g or roughly 50% of one's total calories under normal, non-stressful conditions. This means that, on average, at least half of one's daily energy must be derived from non-protein sources, either fat, oils, or carbohydrates. Protein intakes above this threshold, especially if they fluctuate sharply from day to day, may exceed the rate at which the liver can metabolize high levels of amino acids, and the body can synthesize and excrete urea, leading to impairment of liver and kidney function, as well as other potentially serious health consequences, and perhaps even death (Cahill 1986; McArdle *et al.* 1986; McGilvery 1983; Miller & Mitchell 1982; Speth 1990; Whitney & Hamilton 1984). Clearly, this is a large amount of protein, far in excess of the levels normally seen in the diets of ethnographically documented hunters and gatherers, with the notable exception of the Eskimos, whose all-meat diet often led to protein intakes that approached 40–45% of calories (see Speth (1989) and references therein). Thus, if this were the only limit, it would be of little concern to us in reconstructing early hominid diet, as virtually no one postulates regular per capita protein intakes anywhere near this magnitude some 2.0 to 2.5 Ma BP in the Plio-Pleistocene.

However, there is a small but growing body of evidence suggesting that the safe upper limit to total protein intake for pregnant women may, in fact, be much lower and, if substantiated by further nutritional research, could be much more relevant to our understanding of the food choices of modern hunters and gatherers and perhaps early hominids as well. A woman's protein needs increase during pregnancy, and these requirements must be met in order for her to produce a viable, healthy offspring. Extremely low maternal protein intakes, below about 5–6% of calories, may be detrimental to the health of the foetus (Martorell & Gonzalez-Cossio 1987; National Academy of Sciences 1985; Winick 1989). As maternal protein intakes increase above this minimum threshold, mass at birth and foetal health generally improve. However, several recent studies suggest that supplementation of maternal diets with protein in excess of about 25% of total calories (i.e. above about 100–150 g), even in diets that are otherwise balanced and calorically adequate, may lead to declines, not continued gains, in infant birth mass, and perhaps also to increases in perinatal morbidity and mortality and even cognitive impairment. Premature infants appear to be most vulnerable to high maternal protein supplements (Rush *et al.* 1980; Rush 1982, 1986, 1989; Sloan 1985; Worthington-Roberts & Williams 1989, p. 88).

Birth masses may also decline when the mother's total calorie intake is restricted (Brooke 1987; Lechtig *et al.* 1978; National Academy of Sciences 1985; Wray 1978), but the decline appears to be most extreme when her diet is both low in energy and high in protein (Martorell & Gonzalez-Cossio 1987; Rush 1989; Winick 1989). This is illustrated, for example, by data from Motherwell, a small community in Scotland, where for 30 years pregnant women were advised to consume a comparatively low energy diet consisting of about 1500 kcal and 85 g of protein (*ca.* 23% of total calories). Birth masses of Motherwell infants over this period were, on average, about 400 g lower than those of infants born in Aberdeen during the same period (Kerr-Grieve *et al.* 1979; Winick 1989). This striking decline in average mass at birth rivals the declines seen during wartime famines (Kerr-Grieve *et al.* 1979; Rush 1989).

The relevance of this threshold to the present discussion is that for a major segment of a foraging population – the pregnant women – as much as 70% of their daily calories may have to be obtained from non-protein sources. During much of the year this may pose little or no problem, as women traditionally do most of the plant food collecting in these groups, a division of labour that normally assures them access to ample supplies of carbohydrates and oils (Lee & DeVore 1968). But during seasonal or inter-annual low points in food availability (i.e. the late winter and spring in temperate and northern latitudes or the late dry season and early rainy season in more southerly latitudes), as edible wild plant foods become scarce, unreliable, or more costly in time or effort to procure, the pregnant females in particular may face an increasingly difficult task of maintaining adequate non-protein energy intakes (Speth 1983, 1989, 1990; Speth & Spielmann 1983). These recurrent seasonal or inter-annual resource low points, therefore, may become critical adaptive 'bottlenecks' (Wiens 1977) for hunters and gatherers, and their responses to these hold important clues to understanding the foraging behaviour and food choices of both modern and pre-modern human foragers.

3. UNRESOLVED ISSUES

(a) *Contemporary human foragers*

The preceding section outlined the basic framework of the 'excess protein' issue, looking in particular at the amount of protein that can be safely ingested by pregnant women. But there are many facets of this argument that remain poorly understood, and in this section I specifically focus on these aspects.

First, the very existence of an upper limit to the amount of protein a pregnant women can safely consume on a regular basis remains controversial among nutritionists. Assessing the relationship between dietary protein intake and pregnancy outcome has been compounded by difficulties in obtaining reliable, quantitative data on the amount and nutritional composition of foods consumed by women over the course of their pregnancies, as well as by problems in controlling for differences among subjects, both before and during pregnancy, in factors such as height; body mass; overall health; the use of drugs, alcohol and cigarettes; economic status and level of education. As a consequence, many of the presently available studies suffer from flaws in research design and measurement precision (for a detailed review of prior research, see Rush (1989)).

Second, and perhaps more importantly, the actual mechanisms by which high dietary protein intakes impinge on the health and well-being of the developing foetus remain to be worked out (see Rush 1989; Sloan 1985).

Third, it is not yet clear whether high protein intakes are deleterious to the foetus throughout the course of a pregnancy or only during a particular trimester. Because the developing foetus is particularly susceptible to teratogenic substances during the first trimester, the period of embryonic organogenesis, this may also be the time in a pregnancy when excessive protein intakes would be most problematic. Interestingly, this is also the trimester in which 'pregnancy sickness', as well as intense food aversions and food cravings, are most in evidence and, in cross-cultural studies, meat and meat odours are among the most common and widespread aversions, whereas carbohydrates are among the most common cravings (Dickens & Trethowan 1971; Hook 1978; Profet 1989; Tierson *et al.* 1985). One also finds widespread and seemingly irrational food taboos and inequitable sharing practices among traditional hunter-gatherer and horticultural societies that effectively limit or block pregnant women from access to meat (Aunger 1991; Speth 1990; Spielmann 1989). Unfortunately, however, no one has yet examined these practices to determine whether they are in effect throughout the course of a pregnancy or only during a particular trimester. Despite the many uncertainties, these factors together provide very tentative evidence that high levels of protein may be deleterious to the foetus primarily or exclusively during the first trimester of pregnancy.

Fourth, if high protein intakes are, in fact, detrimental to pregnancy outcome, how did pregnant women in traditional Eskimo and other northern latitude foraging societies cope with this problem? These peoples commonly consumed average daily protein intakes in excess of 25% of total calories and levels on the order of 40–45% were not uncommon (e.g. Draper 1977; see also discussion and references in Speth (1989, 1990)). What did pregnant women in arctic and subarctic environments consume to avoid high protein intakes? Unfortunately, the existing dietary and nutritional literature for these groups is extremely sketchy, providing little quantitative data on diet broken down by sex, and none according to reproductive status. As a consequence, I can offer only a few tentative suggestions here that may serve as working hypotheses for future nutritional research among these groups. For example, pregnant women probably consumed greater proportions of fat in their diet than did other adults (H. V. Kuhnlein, personal communication). They also appear to have augmented the meagre carbohydrate component of their diet by collecting berries and other terrestrial plant foods (see, for example, Eidlitz (1969); Giffen (1930); Nickerson *et al.* (1973)), as well as kelp washed up on the beach or harvested through the ice in winter, by extracting the fat-rich fly larvae from the hides of caribou and reindeer, and by consuming the partly digested stomach and rumen contents of mammalian and avian herbivores (H. V. Kuhnlein, personal communication; Eidlitz 1969).

The fifth and final issue concerns the significance of the protein threshold in terms of its actual or potential demographic impact on foragers. What is the relationship between excessive protein consumption by a pregnant woman, particularly during periods of reduced overall energy intake, and the level of perinatal morbidity and mortality? Anthropologists have long suspected that seasonally marginal calorie intakes play an important role in the low completed fertility of extant foraging populations such as the Kalahari San or Bushmen and the Australian Aborigines (see, for example, Howell (1979, 1986); Lager & Ellison (1987)). The arguments presented here suggest that high dietary protein intakes during these seasonal 'bottlenecks' may also play a role.

(*b*) *Living higher primates*

Up to this point, I have been concerned with the existence and nature of a dietary protein threshold among pregnant women in contemporary human foraging populations. Obviously, even if the threshold proves to be both real and demographically significant in modern humans, this in no way shows that such a threshold also existed in early hominids. This is clearly a much more difficult problem to deal with. Although less than ideal, a common way of approaching this sort of problem is to see if an analogous phenomenon exists in our closest living primate relatives, the chimpanzees. If the answer is yes, this would suggest that a protein threshold was also present in the common ancestor of both chimpanzees and modern humans and hence probably also in Plio-Pleistocene hominids. Thus, the next step is to raise the same issues concerning chimpanzees that were raised above for contemporary foragers: do female chimpanzees face seasonal or inter-annual shortfalls in resource availability and total energy intake (see, for example, Wrangham (1975, 1977) for evidence of substantial seasonal body mass fluctuations in chimpanzees)? Does the absolute amount and proportional contribution of protein in their diet increase during these stressful periods, as their principal fruits (generally comparatively low-protein foods) become less abundant or accessible (e.g.

Wrangham 1975)? Is there any evidence for the existence of a protein threshold in pregnant chimpanzee females above which pregnancy outcome is negatively impacted? If so, what proportion of total calories can safely come from protein? Does the threshold persist over the entire course of the pregnancy or only during a particular trimester? What, if any, is the potential demographic impact of excessive protein intakes? And how do female chimpanzees cope behaviourally and dietarily with the threshold? These are issues clearly in need of explicit investigation by specialists in primate and human nutrition.

4. SUMMARY AND CONCLUSIONS

This paper raises many more questions than it answers. To summarize briefly, there is a growing body of nutritional evidence to suggest that pregnant women may face a dietary constraint, at least during the first trimester of pregnancy, in which their protein intake (plant and animal combined) must be kept below about 25–30% of their total daily energy intake, to protect the health and well-being of the developing foetus. This constraint may be most critical to forager women during seasonal or inter-annual periods of reduced overall food availability. For most modern human populations, particularly those in western industrial nations, protein intakes rarely approach this threshold and it is probably therefore of little or no significance. Among ethnographically and ethnohistorically documented foragers, however, protein intakes approaching and even exceeding this level are actually fairly common, especially during the late winter – spring in temperate and northern latitudes and the late dry season – early rainy season in the tropics and sub-tropics. At such times, foragers are faced with reduced overall energy intakes and they may be forced to rely increasingly on foods, such as seeds, nuts and meat, that are protein-rich but often low in fat. For pregnant females, such times may place the developing foetus at increased risk. This paper basically seeks further input from nutritionists and primatologists to help answer two basic questions: is this protein threshold real and demographically significant in modern human foraging populations, and does a similar threshold affect pregnant female chimpanzees? If the answer to both of these questions is yes, we can then begin to explore systematically the implications such a threshold might have for the diet and foraging strategies of early hominids.

I thank Dr E. Widdowson and Dr A. Whiten for their helpful comments in preparing the final manuscript for publication. The paper was prepared while I was a Weatherhead Resident Scholar at the School of American Research (SAR) in Santa Fe, New Mexico. I am grateful to the SAR for providing the support and intellectual environment in which to explore and develop the ideas presented here.

REFERENCES

Aunger, R. 1991 The nutritional consequences of food taboos in the Ituri Forest, Zaire. Paper presented at the annual meeting of the American Association of Physical Anthropologists, Milwaukee, Wisconsin, April 1991.

Binford, L. R. 1978 *Nunamiut ethnoarchaeology*. New York: Academic Press.

Binford, L. R. 1981 *Bones: ancient men and modern myths*. New York: Academic Press.

Blumenschine, R. J. 1987 Characteristics of an early hominid scavenging niche. *Curr. Anthrop.* **28**, 383–407.

Brooke, O. G. 1987 Nutritional requirements of low and very low birth weight infants. *A. Rev. Nutr.* **7**, 91–116.

Bunn, H. T. 1986 Patterns of skeletal representation and hominid substance activities of Olduvai Gorge, Tanzania, and Koobi Fora, Kenya. *J. hum. Evol.* **15**, 673–690.

Bunn, H. T., Bartram, L. E. & Kroll, E. M. 1988 Variability in bone assemblage formation from Hadza hunting, scavenging, and carcass processing. *J. anthrop. Archaeol.* **7**, 412–457.

Cahill, G. F. Jr. 1986 The future of carbohydrates in human nutrition. *Nutr. Rev.* **44**, 40–43.

Dickens, G. & Trethowan, W. H. 1971 Cravings and aversions during pregnancy. *J. Psychosomatic Res.* **15**, 259–268.

Draper, H. H. 1977 The aboriginal Eskimo diet in modern perspective. *Am. Anthrop.* **79**, 309–316.

Eidlitz, K. 1969 *Food and emergency food in the circumpolar area.* Studia Ethnographica Upsaliensia **32**. Uppsala, Sweden: Almqvist & Wiskell.

Giffen, N. M. 1930 *The roles of men and women in Eskimo culture.* University of Chicago Press.

Gordon, K. D. 1987 Evolutionary perspectives on human diet. In *Nutritional anthropology* (ed. F. E. Johnston), pp. 3–39. New York: Alan R. Liss.

Hill, K. 1982 Hunting and human evolution. *J. hum. Evol.* **11**, 521–544.

Hook, E. B. 1978 Dietary cravings and aversions during pregnancy. *Am. J. clin. Nutr.* **31**, 1355–1362.

Howell, N. 1979 *Demography of the Dobe !Kung.* New York: Academic Press.

Howell, N. 1986 Feedbacks and buffers in relation to scarcity and abundance: studies of hunter-gatherer populations. In *The state of population theory: forward from Malthus* (ed. D. Coleman & R. Schofield), pp. 156–187. Oxford: Basil Blackwell.

Isaac, G. Ll. & Crader, D. 1981 To what extent were early hominids carnivorous? In *Omnivorous primates: gathering and hunting in human evolution* (ed. R. S. O. Harding & G. Teleki), pp. 37–103. New York: Columbia University Press.

Kerr-Grieve, J. F., Campbell-Brown, B. M. & Johnstone, F. D. 1979 Dieting in pregnancy: a study of the effect of a high protein low carbohydrate diet on birthweight on an obstetric population. In *International colloquium on carbohydrate metabolism in pregnancy and the newborn 1978* (ed. H. W. Sutherland & J. M. Stowers), pp. 518–534. New York: Springer-Verlag.

Klein, R. G. 1989 *The human career: human biological and cultural origins.* University of Chicago Press.

Lager, C. & Ellison, P. T. 1987 Effects of moderate weight loss on ovulatory frequency and luteal function in adult women. *Am. J. phys. Anthrop.* **72**, 221–222.

Lechtig, A., Delgado, H., Martorell, R., Richardson, D., Yarbrough, C. & Klein, R. E. 1978 Effect of maternal nutrition on infant mortality. In *Nutrition and human reproduction* (ed. W. H. Mosley), pp. 147–174. New York: Plenum Press.

Lee, R. B. & DeVore, I. (eds). 1968 *Man the hunter.* Chicago: Aldine.

McArdle, W. D., Katch, F. I. & Katch, V. L. 1986 *Exercise physiology: energy, nutrition, and human performance*, p. 545. Philadelphia: Lea & Febiger.

McGilvery, R. W. 1983 *Biochemistry: a functional approach*, 3rd ed. Philadelphia: W. B. Saunders.

Martorell, R. & Gonzalez-Cossio, T. 1987 Maternal nutrition and birth weight. *Ybk. phys. Anthrop.* **30**, 195–220.

Miller, S. A. & Mitchell, G. V. 1982 Optimisation of human protein requirements. In *Food proteins* (ed. P. F. Fox & J. J. Condon), pp. 105–120. London: Applied Science Publishers.

National Academy of Sciences. 1985 *Preventing low birthweight.* Washington, D.C.: Committee to Study the Prevention of Low Birthweight, Division of Health Promotion and Disease Prevention, Institute of Medicine, National Academy Press.

Nickerson, N. H., Rowe, N. H. & Richter, E. A. 1973 Native plants in the diets of north Alaskan Eskimos. In *Man and his foods* (ed. C. E. Smith, Jr), pp. 3–27. University of Alabama Press.

O'Connell, J. F., Hawkes, K. & Blurton Jones, N. 1988 Hadza scavenging: implications for Plio-Pleistocene hominid subsistence. *Curr. Anthrop.* **29**, 356–363.

Peters, C. R., O'Brien, E. M. & Box, E. O. 1984 Plant types and seasonality of wild-plant foods, Tanzania to southwestern Africa: resources for models of the natural environment. *J. hum. Evol.* **13**, 397–414.

Potts, R. 1982 Lower Pleistocene site formation and hominid activities at Olduvai Gorge, Tanzania. Ph.D. thesis, Harvard University.

Profet, M. 1989 Pregnancy sickness as adaptation: a deterrent to maternal ingestion of teratogens. In *The adapted mind: evolutionary psychology and the generation of culture* (ed. J. Barkow, L. Cosmides & J. Tooby). (Unpublished manuscript.)

Rush, D. 1982 Effects of changes in protein and calorie intake during pregnancy on the growth of the human fetus. In *Effectiveness and satisfaction in antenatal care* (ed. M. Enkin & I. Chalmers), pp. 92–113. Clinics in Developmental Medicine 81/82. Philadelphia: J. B. Lippincott.

Rush, D. 1986 Nutrition in the preparation for pregnancy. In *Pregnancy care: a manual for practice* (ed. G. Chamberlain & J. Lumley), pp. 113–139. New York: John Wiley & Sons.

Rush, D. 1989 Effects of changes in protein and calorie intake during pregnancy on the growth of the human fetus. In *Effective care in pregnancy and childbirth*, vol. 1 (*Pregnancy*) (ed. I. Chalmers, M. Enkin & M. J. N. C. Keirse), pp. 255–280. Oxford University Press.

Rush, D., Stein, Z. & Susser, M. 1980 A randomized controlled trial of prenatal nutritional supplementation in New York City. *Pediatrics* **65**, 683–697.

Sept, J. M. 1984 Plants and early hominids in East Africa: a study of vegetation in situations comparable to early archaeological site locations. Ph.D. thesis, University of California.

Shipman, P. 1983 Early hominid lifestyle: hunting and gathering or foraging and scavenging? In *Animals and archaeology*, vol. 1. (*Hunters and their prey*) (ed. J. Clutton-Brock & C. Grigson), pp. 31–49 (British Archaeological Reports (International Series) **163**). Oxford.

Sloan, N. L. 1985 Effects of maternal protein consumption on fetal growth and gestation. Ph.D. thesis, Columbia University.

Speth, J. D. 1983 *Bison kills and bone counts: decision making by ancient hunters.* University of Chicago Press.

Speth, J. D. 1987 Early hominid subsistence strategies in seasonal habitats. *J. archaeol. Sci.* **14**, 13–29.

Speth, J. D. 1989 Early hominid hunting and scavenging: the role of meat as an energy source. *J. hum. Evol.* **18**, 329–343.

Speth, J. D. 1990 Seasonality, resource stress, and food sharing in so-called 'egalitarian' foraging societies. *J. anthrop. Archaeol.* **9**, 148–188.

Speth, J. D. & Spielmann, K. A. 1983 Energy source, protein metabolism, and hunter-gatherer subsistence strategies. *J. anthrop. Archaeol.* **2**, 1–31.

Spielmann, K. A. 1989 A review: dietary restrictions on hunter-gatherer women and the implications for fertility and infant mortality. *Hum. Ecol.* **17**, 321–345.

Teleki, G. 1981 The omnivorous diet and eclectic feeding habits of chimpanzees in Gombe National Park, Tanzania. In *Omnivorous primates: gathering and hunting in human evolution* (ed. R. S. O. Harding & G. Teleki), pp. 303–343. New York: Columbia University Press.

Tierson, F. D., Olsen, C. L. & Hook, E. B. 1985 Influence of cravings and aversions on diet in pregnancy. *Ecol. Food Nutr.* **17**, 117–129.

Whitney, E. N. & Hamilton, E. M. N. 1984 *Understanding nutrition*, 3rd edn, p. 145. St Paul, Minnesota: West Publishing Company.

Wiens, J. A. 1977 On competitive and variable environments. *Am. Sci.* **65**, 590–597.

Winick, M. 1989 *Nutrition, pregnancy, and early infancy.* Baltimore: Williams & Wilkins.

Winterhalder, B. & Smith, E. A. (eds). 1982 *Hunter-gatherer foraging strategies: ethnographic and archaeological analyses.* University of Chicago Press.

Worthington-Roberts, B. & Williams, S. R. 1989 *Nutrition in pregnancy and lactation*, 4th edn. St Louis: Times Mirror/Mosby.

Wrangham, R. W. 1975 The behavioural ecology of chimpanzees in Gombe National Park, Tanzania. Ph.D. thesis, University of Cambridge.

Wrangham, R. W. 1977 Feeding behaviour of chimpanzees in Gombe National Park, Tanzania. In *Primate ecology: studies of feeding and ranging behaviour in lemurs, monkeys and apes* (ed. T. H. Clutton-Brock), pp. 503–538. London: Academic Press.

Wray, J. D. 1978 Maternal nutrition, breast-feeding and infant survival. In *Nutrition and human reproduction* (ed. W. H. Mosley), pp. 197–229. New York: Plenum Press.

Discussion

E. M. Widdowson (*9 Boot Lane, Barrington, Cambridge, U.K.*). Dr Speth's paper raises some interesting questions about protein metabolism. We did an experiment on rats many years ago (Cabak *et al.* 1963) in which some were fed a low-protein diet in measured amounts, but they did not eat enough of it to maintain body mass. Others had a high-protein diet (48% protein) in restricted amounts so that their mean mass remained the same as those on the low-protein diet. A third group had a normal rat diet and they served as the controls. A third of the animals having the high-protein, restricted-energy diet died during the experiment, but none of the others. Those that remained had high concentrations of urea and low concentrations of glucose in their blood. Does Dr Speth think a high-protein diet exerts a harmful effect primarily when the energy intake is restricted?

Reference

Cabak, V., Dickerson, J. W. T. & Widdowson, E. M. 1963 Response of young rats to deprivation of protein or of calories. *Brit. J. Nutr.* **17**, 601–616.

O. T. Oftedal (*Smithsonian Institution, Washington D.C., U.S.A.*). I agree with Dr Widdowson that the relationship between protein excess and energy requirements may be particularly important. The examples of polar and Western explorers that Dr Speth cites were often under conditions of extreme energy demand due to cold exposure or severe physical exhaustion. Under such circumstances, low-fat diets (i.e. lean tissues) may well fail to meet energy demands despite deamination and catabolism of amino acids as an energy source. In subfreezing conditions explorers might also have difficulty obtaining enough melted water to sustain urea excretion via urine formation. I suspect that primates in arid conditions that do not have access to drinking water may be limited in their ability to use high protein foods due to water shortage. Dr Speth might also want to examine the literature on high-protein diets for dogs as this has been somewhat of a controversial issue, at least in the United States.

J. D. Speth. All of the studies I am aware of point to the combination of high dietary protein intake together with low total energy intake as potentially the most problematic situation. Thus, in the case of human foragers or hunter-gatherers, high protein intakes are most likely to pose a problem primarily at times of year when their total food intake is restricted, in other words, during the winter and spring in more northerly latitudes or during the dry season in the tropics and subtropics. At such times the most vulnerable segment of the population may be the pregnant women.

R. A. Foley (*Department of Biological Anthropology, University of Cambridge, U.K.*). In view of the fact that the occurrence of 'protein starvation' seems specific to high-latitude populations and individuals in periods of winter, could it be that this phenomenon is associated with high energy expenditure due to the thermoregulatory problems of living at very cold temperatures?

J. D. Speth. Although 'protein starvation' has been mentioned most commonly and explicitly in the ethnographic and explorer literature of the northern latitudes, the phenomenon is by no means restricted just to these parts of the world. For example, by the end of the dry season in the tropics and subtropics, foragers may be faced with marginal or inadequate total energy intakes, perhaps exacerbated by heavy work loads, forcing them to obtain the bulk of their energy from fat-depleted game. Under these circumstances, high protein intakes pose the same problem they do in the arctic and subarctic. In fact, shortage of water may make the problem even more acute.

P. Van Soest (*324 Morrison Hall, Cornell University, New York, U.S.A.*). High protein consumption had been a feature of some dietary fads in the United States designed to promote weight loss. This is regarded as hazardous by many doctors who point out the dangers of uremic poisoning in subjects with liver or kidney disease. High-protein diets may mean diets low in both carbohydrates or fat. Low carbohydrate plus fat sets the need for gluconeogenesis from amino acids to maintain body glucose needs. The production of NH_3 from this process plus that from protein used for energy may overload the liver capacity for urea synthesis. NH_3 is extraordinarily toxic to tissue cells. One theory of colon cancer is based on this concept. The cost of energy to detoxify one gram of ammonic nitrogen is about 45 kJ. This is a high energy cost. Ruminants are very sensitive to overfeeding of protein.

The calcium-phosphorous ratio of meat is about 1:15, whereas requirements are 2:1. Thus meat is grossly deficient in calcium and excessive in phosphorus. A nutrition problem for lions and cats in zoos, and dogs fed high-meat diets, is the occurrence of rickets.

Human dietary change

STANLEY J. ULIJASZEK

Department of Biological Anthropology, University of Cambridge, Downing Street, Cambridge CB2 3DZ, U.K.

SUMMARY

The transition from hunting and gathering to agriculture and animal husbandry in the Near East and Mediterranean Region began some 12000 years ago. The ecological changes associated with this change are known to have been related to higher levels of stress from undernutrition and infectious disease. Certain pathologies found in human skeletal remains from this time are indicative of anaemia and osteoporosis, although it is not clear whether they had clear nutritional aetiologies. In this paper, dietary changes associated with changes in subsistence practices in this region are described. In addition, quantitative modelling of possible patterns of dietary and nutrient intakes of adult males before, and soon after, the establishment of agrarian economies is used to examine the proposition that the skeletal pathologies porotic hyperostosis, cribra orbitalia and porotic hyperostosis may have been due to nutritional deficiencies. The results suggest that protein deficiency was only likely if subjects were suffering from chronic energy deficiency (CED) and their diet contained no meat. Dietary calcium deficiency was possible after the transition to cultivation and animal husbandry, in the presence of moderate or severe CED. Anaemias, although present after the transition, were unlikely to have had dietary aetiologies, regardless of the severity of CED.

1. INTRODUCTION

The transition from hunting and gathering to agriculture and animal husbandry began in the Near East and Mediterranean Region some 12000 years before present (BP), with the deliberate growing of wild cereal crops and the taming of small mammals as means of expanding the food supply in response to population increase (Hillman *et al.* 1989). The Neolithic (10500–5750 years BP) was the period in prehistory in which a pattern of village settlement based on subsistence farming and stockbreeding became the basis of existence for communities throughout the Near East (Moore 1985), and was associated with greater physiological stress due to undernutrition and infectious disease (Cohen 1989).

Although there is a vast potential corpus of data about prehistoric human nutritional pathology, there are problems of interpretation (Armelagos 1987). The modelling of possible dietary and nutrient intakes under a variety of conditions provides an alternative way of examining and augmenting our knowledge of the nutritional ecology of past populations. In this paper, such a scheme is used to examine the proposition that the aetiologies of the skeletal pathologies, porotic hyperostosis, cribra orbitalia and osteoporosis found in human remains from the Near East and Mediterranean regions before and during the Neolithic, were due to nutritional deficiencies.

2. CHANGES IN SUBSISTENCE PRACTICES IN THE NEAR EAST AND THE MEDITERRANEAN

Locations in the Near East and Mediterranean at which diet and subsistence before and during the Neolithic have been studied include sites in the Jordan Rift Valley, Western Turkey, Israel, Iran, Syria and Iraq. Evidence for dietary usage comes from study of assemblages of animal bones, plant food and faecal remains at, or close to, sites of human habitation (Harris & Hillman 1989; Clutton-Brock 1989).

Between about 20000 and 12000 years BP the region provided a mosaic of so-called 'optimal' habitats in which semi-nomadic groups practised hunting and gathering (Bar-Yosef 1987). These 'optimal' habitats were separated by less favourable habitats that had

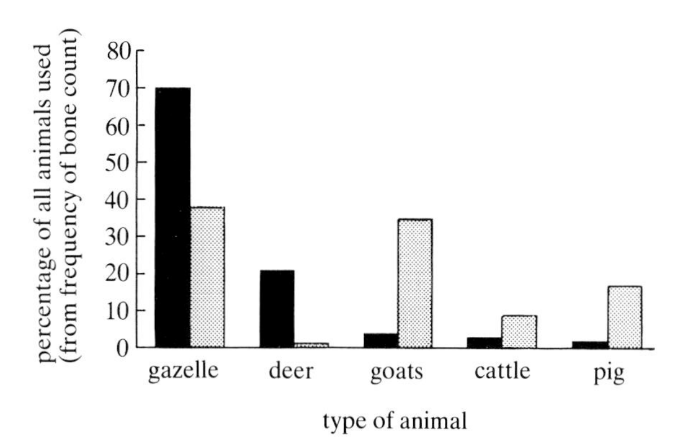

Figure 1. Ungulate faunal spectra in pre-Neolithic times (solid bars) and in the Neolithic (shaded bars). From Smith *et al.* (1984).

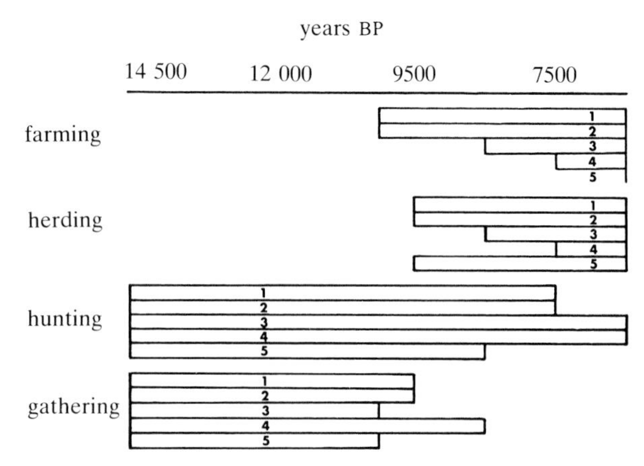

Figure 2. Tentative scheme of major subsistence activities in the Southern Levant. Regions: 1, Galilee, Mount Carmel, Judean Hills, Coastal Plain; 2, Jordan Valley; 3, Western Negev, Northern Sinai; 4, Southern Sinai; 5, Negev Highlands. After Smith *et al.* (1984).

lower carrying capacities and population densities of hunter–gatherers (Flannery 1969). Hunting of hoofed mammals including gazelle, antelope and deer was widespread (Flannery 1965; Jacobsen 1969), although the hunting or gathering of fish, crabs, molluscs and snails, partridges and migratory waterfowl increased with time (Flannery 1969). Gathered vegetable sources of food included root plants, wild pulses, almonds, pistachio and hazel nuts and fruits such as apples (Angel 1984). Wild cereal grains were also gathered, and these included wild strains of wheat and barley (Reed 1977).

By 11000 years BP wild sheep and goats were being domesticated (Butzer 1971; Reed 1977), and the domestication of cattle and pigs followed soon after (Angel 1984). The planting of wheat and barley was well underway before 9000 BP (Hopf 1969; Van Zeist & Bakker-Heeres 1979) as was that of pulses. The exploitation of fruits and nuts also shifted into systematic cultivation. Associated with this was a shift in the extent to which different species of animal were used as food (figure 1), and in an enormous reduction in the extent of meat consumption (Angel 1984).

The shift from hunting and gathering to cultivation and domestication was not an immediate and direct one; during the transition human groups probably varied in the extent to which they practised each strategy. In addition, the time at which farming and herding took precedence over hunting and gathering varied from site to site (figure 2), although by 7000 years BP the transition from one mode of subsistence to another was more or less complete.

3. PALAEOPATHOLOGIES AT THE ORIGINS OF AGRICULTURE

Evidence in support of the view that the transition from hunting and gathering to cultivation and animal husbandry in the Near East and Mediterranean was associated with an increase in disease and nutritional stress comes from various analyses of skeletal remains (Angel 1984; Smith *et al.* 1984). Estimates of the stature of adult males from femoral size by using the equations of Trotter & Gleser (1952) indicate a decline from an average of about 180 cm to about 170 cm between 32000–11000 years BP and 11000–9000 years BP (Angel 1984). This decline has been attributed to generalized undernutrition (Angel 1984), although other factors, such as increased stress from infectious disease in the more densely populated settlements at this time, or inbreeding in isolated communites might be equally important.

Specific bone pathologies which are related to nutritional or disease stress and have been identified in skeletal material from the region include porotic hyperostosis and cribra orbitalia. Porotic hyperostosis can be recognized by a thickened, porous, sievelike appearance of certain parts of the skeleton. Cribra orbitalia is porotic hyperostosis of the orbital surfaces of the skull. Both pathologies are diagnostic of anaemia, although it is not possible to give a specific aetiology on this basis alone (Huss-Ashmore *et al.* 1982).

Porotic hyperostosis has been reported to be present in skeletal remains from the Neolithic onward (Rathbun 1984). Although a primary dietary aetiology has been proposed (Moseley 1965), Stuart-Macadam (1991) has recently suggested that porotic hyperostosis in prehistoric populations is more likely to have been due to continuous or heavy pathogen loads than to iron deficiency. Cribra orbitalia may be due to dietary deficiency of iron or ascorbic acid, or to any infection which leads to chronic anaemia (Wing & Brown 1979). Its presence in the Neolithic Near East has been attributed to the effects of infectious disease on the bioavailability and utilization of iron rather than to dietary deficiency per se (Smith *et al.* 1984).

Osteoporosis specifically due to dietary calcium deficiency is rare in the modern world, but the possibility that it may have been more common in prehistoric populations cannot be discounted. In contemporary populations, osteoporosis is a condition where the bone density of an individual is lower than might be expected for their age, and may occur in the elderly population, as well as adults who are extremely inactive, or in women who are amenorrheic due to very high levels of physical activity (British Nutrition Foundation Task Force on Calcium 1989). Osteoporosis in skeletal material from past populations has often been ascribed to calcium deficiency by palaeoanthropologists, even though there is only a poor physiological basis on which to base this conclusion. Indeed, Kurth & Rohrer-Ertl (1981) found that long bones of 28 individuals from the Neolithic period in the Near East showed signs of osteoporosis, which they attributed to calcium deficiency. However, there is scant evidence of osteoporosis in skeletal material from earlier times (Smith *et al.* 1984).

Neolithic Near Eastern populations are known to have had high prevalences of anaemia and osteoporosis, but it is not clear whether these pathologies had specific dietary aetiologies, or were related to other factors. One way of evaluating the relative importance of dietary factors in the aetiology of these pathologies is to model dietary and nutrient intakes before, and during, the Neolithic, and to relate the latter to current

Table 1. *Physical activity levels for adult males engaged in a variety of subsistence practices*

group or district	country	subsistence type	physical activity level (TEE/BMR)	reference
Ache	Paraguay	hunter–gatherer	2.08[a]	Hill *et al.* (1985)
Machiguenga	Peru	hunter–horticulturalist	2.09[b]	Montgomery & Johnson (1974)
Wopkaimin	Papua New Guinea	hunter–horticulturalist	1.65[b,c]	S. J. Ulijaszek & T. Brown (unpublished results)
Pari	Papua New Guinea	hunter–horticulturalist	1.42	Hipsley & Kirk (1962)
	Guatamala	maize cultivation	2.32	Viteri *et al.* (1971)
	Philippines	rice cultivation	2.25	de Guzman *et al.* (1974)
Varanin	Iran	wheat cultivation	2.10[d]	Brun *et al.* (1979)
	Burma	rice cultivation	2.02[b]	Tin-May-Than & Ba-Aye (1985)
	Gambia	rice and peanut cultivation	2.02[b]	Fox (1953)
Tamil Nadu	India	rice cultivation	2.00[d]	McNeill *et al.* (1987)
Sundanese	Indonesia	rice cultivation	1.96[a]	Suzuki (1988)
	Burkina Faso	millet cultivation	1.89[b]	Brun *et al.* (1981)
	Ivory Coast	mixed cultivation	1.68[a]	Dasgupta (1977)
Lufa	Papua New Guinea	sweet potato cultivation	1.64[a]	Norgan *et al.* (1974)
	Uganda	maize and plantain cultivation	1.63[a]	Cleave (1970)
	Cameroun	millet cultivation	1.54[a]	Guet (1960)
Kaul	Papua New Guinea	taro and plantain cultivation	1.52[a]	Norgan *et al.* (1974)

[a] From estimates of BMR from body mass, and total energy expenditure from activity diaries.
[b] Average of two seasons.
[c] Estimate of total energy expenditure from activity diaries.
[d] Average of four seasons.

estimates of nutritional requirements (FAO/WHO 1974, FAO/WHO/UNU 1985).

4. MODELLING CHANGES IN NUTRIENT INTAKE

In this paper, model pre-Neolithic and Neolithic men are created, based on knowledge of stature at those times, and on current understanding of energy expenditures of contemporary hunter–gatherers and hunter–horticulturalists, and simple cultivators. Assuming that these men are in energy balance, intakes of protein, iron and calcium are calculated from dietary reconstructions based on evidence from the literature. Intakes of these nutrients are then estimated for the model Neolithic man, assuming that he is suffering from chronic energy deficiency (CED) at various levels of severity.

Dietary energy requirements are related to energy expenditure, which in turn is related to activity level and to basal metabolic rate (BMR) (FAO/WHO/UNU 1985). It is possible to express total daily energy expenditure (TEE) as a multiple of BMR (Waterlow 1986). In this way, differences in TEE due to differences in body size can be controlled for, giving a general measure of activity called the physical activity level (PAL) (James *et al.* 1988). Table 1 gives PALs obtained by a variety of means, for contemporary populations practising a number of subsistence strategies.

Hunter–gatherers and hunter–horticulturalists have PALs that range from 1.42 to 2.09 whereas the PAL of cultivators ranges from 1.52 to 2.32. On average, the difference in energy expenditure due to physical activity between groups practising these different types of subsistence is small, and it can be postulated that the switch from hunting and gathering to cultivation and animal husbandry involved little change in PAL. In the model to be elaborated, the pre-Neolithic man is assigned a PAL of 1.9, and the Neolithic man, 2.0.

Seasonal variation in activity levels, energy expenditure and therefore energy requirement almost certainly became exaggerated with the onset of agriculture. However, the pathologies with which this paper is concerned are due to chronic rather than acute stress, and seasonal variation in those stresses is unlikely to be of greater significance than the overall stress levels across the entire year.

Body mass index (BMI) is a crude index of body physique which has been recommended for use in the estimation of nutritional status in adults (James *et al.* 1988). BMI is body mass divided by height squared, and a scheme whereby this measure is used in association with PAL is given in table 2.

It is possible to model energy requirements of adult males in pre-Neolithic and Neolithic times at different

Table 2. *Body mass index and physical activity level* (PAL) *as markers for chronic energy deficiency* (CED) *in adults*

(From James *et al.* (1988).)

BMI	PAL[a]	presumptive diagnosis
> 18.5	—	normal
17.0–18.5	> 1.4	normal
	< 1.4	CED grade I
16.0–17.0	> 1.4	CED grade I
	< 1.4	CED grade II
< 16.0	—	CED grade III

[a] PAL = physical activity level (TEE/BMR).

Table 3. *Estimated mass, basal metabolic rate and total energy expenditure of adult males in pre-Neolithic times and the Neolithic*

	pre-Neolithic	Neolithic
height/cm	180	170
BMI/(kg m^{-2})	20.5	20.5
mass/kg	66.4	59.2
BMR/(MJ d^{-1})	7.08	6.63
TEE/(MJ d^{-1})	13.45	13.26

levels of energy nutritional status, assuming that subjects are in energy balance at different levels of BMI and PAL. Data from a variety of contemporary populations show that for hunter–gatherers and hunter–horticulturalists, BMIs of adult males range from 19.0 to 20.7 kg m^{-2}, whereas for simple cultivators, they range from 18.3 to 21.3 kg m^{-2}, respectively. In this analysis, both pre-Neolithic and Neolithic model men are assigned a BMI of 20.5 kg m^{-2}.

Body masses of the model men were calculated from their estimated heights and their assumed BMIs, and used to calculate BMR (Schofield 1985). The TEE of both men was then calculated from their respective BMR and PAL; daily energy requirement was assumed to be equal to energy expenditure. Estimates of BMR and TEE of adult males, based on assumed stature of 180 cm pre-Neolithic and 170 cm in the Neolithic, and on BMI of 20.5 kg m^{-2} at both times, are given in table 3.

TEE and daily energy requirements of both men are very similar, although their diets are likely to have been quite different. Qualitative descriptions of dietary change across the transition from hunting and gathering to agriculture and animal husbandry are numerous (Flannery 1965, 1969; Jacobsen 1969; Reed 1977; Angel 1984), but quantification is rather difficult. Estimates of the relative contribution of different animal species to the diet are far more accurate than those of different plant species, as existing methods in archaeology are more able to quantify the consumption of animal foods than plant foods. Angel (1971) has suggested that meat eating in the Neolithic may have been only 10–20 % of the Palaeolithic norm. This may well have resulted in a decline in the contribution of animal foods to the diet from a level greater than 20 % of the total daily energy intake, to below 10 %.

There is little evidence of the relative contribution of different plant types to the diet of pre-Neolithic human groups. Despite this difficulty, an attempt has been made to estimate the proportion of dietary energy supplied by different plant and animal types, based on the descriptions of Flannery (1969), Jacobsen (1969), Reed (1977) and Angel (1984) (figure 3). It is assumed that wild grasses contributed half of the daily energy intake from plant sources in pre-Neolithic times, and

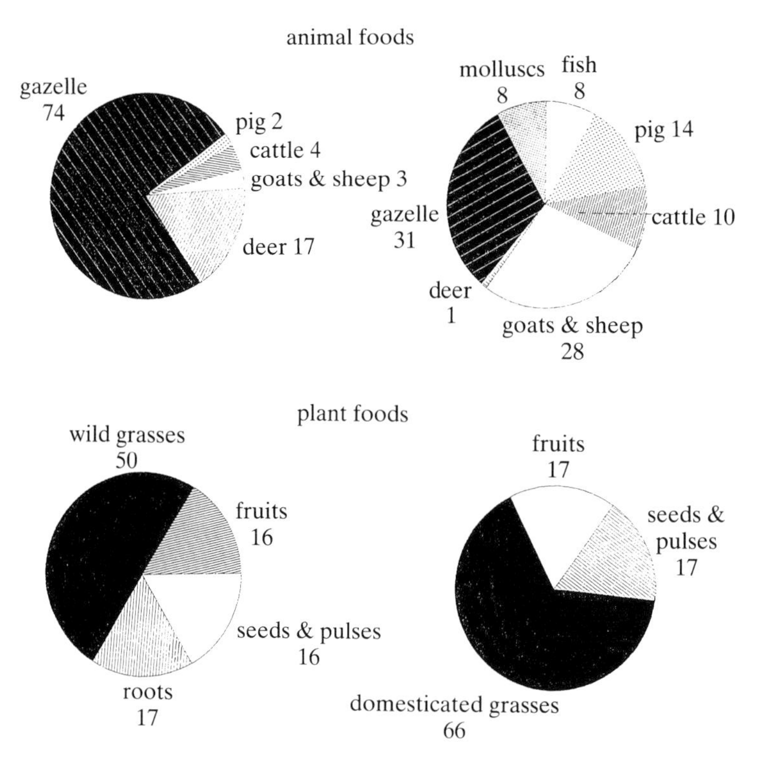

Figure 3. Proportion of total energy intake from different plant and animal sources.

Table 4. *Nutrient composition of animal and vegetable foods used in pre-Neolithic times and the Neolithic*

	nutrient composition (per 100 grams edible portion)			
	energy/kJ	protein/g	iron/mg	calcium/mg
animal foods				
pre-Neolithic	530	16.7	2.8	10
Neolithic	570	16.4	1.6	7
plant foods				
pre-Neolithic	310	2.8	0.7	9
Neolithic	400	3.2	0.4	7

that domesticated grasses (wheat and barley) supplied two thirds of plant food energy intake in the Neolithic. Although recent evidence suggests that the contributions of such grasses may have been considerably lower than this (G. C. Hillman, personal communication), adjusting the figures does not appreciably affect the analysis.

It is also assumed that milk does not contribute to the diet, since it is believed that domesticated animals were not widely used to produce milk for human consumption before the end of the Neolithic period (Bokonyi 1969).

The nutrient composition of edible portions of the animal and vegetable components of the diets at these two times, as estimated from food composition tables (FAO 1982; Paul & Southgate 1978) are given in table 4.

Although there is no direct evidence of the relative contributions of animal and plant sources of food to total energy intake at the periods in question, contemporary hunter–gatherers obtain between 12 and 86% of their dietary energy from animal sources (Hill

Table 5. *Estimates of daily nutrient intakes at varying levels of animal food intake*

proportion of dietary energy from animal sources (%)	daily intake of nutrients: energy density (g/MJ^{-1})	energy/MJ	protein/g	PER[a] (%)	iron/mg	calcium/mg
pre-Neolithic						
80	206	13.45	384.6	47.9	64.7	271
35	258	13.45	267.0	33.3	49.7	325
20	282	13.45	196.0	24.4	42.6	350
15	291	13.45	172.5	21.5	40.4	359
Neolithic						
15	235	13.26	160.7	20.3	20.1	217
10	240	13.26	143.7	18.1	17.2	223
5	245	13.26	125.5	15.8	14.9	228
0	250	13.26	106.7	13.4	13.3	232

[a] Protein energy ratio is the proportion of dietary energy from protein.

Table 6. *Estimates of daily nutrient intakes in adult males in the Neolithic period, assuming different levels of chronic energy deficiency*

	physical activity level					
	2.0 × BMR			1.35 × BMR		
body mass index (kg m^{-2})	protein/g	iron/mg	calcium/mg	protein/g	iron/mg	calcium/mg
proportion of dietary energy from animal sources = 0%						
18.0	98.8	12.4	216	66.6	8.3	146
16.5	94.4	11.8	197	63.8	8.0	139
15.0	90.1	11.2	191	60.8	7.6	133
proportion of dietary energy from animal sources = 5%						
18.0	116.7	13.9	211	78.7	9.4	143
16.5	111.5	13.2	202	75.3	9.0	137
15.0	106.4	12.7	193	71.8	8.6	130
proportion of dietary energy from animal sources = 10%						
18.0	133.8	15.4	208	90.3	10.4	140
16.5	127.9	14.7	198	86.4	10.1	136
15.0	122.0	14.0	189	82.4	9.5	128
proportion of dietary energy from animal sources = 15%						
18.0	149.6	18.7	202	100.9	12.6	136
16.5	143.0	17.9	193	96.5	12.1	130
15.0	136.3	17.1	184	92.1	11.5	124

1982). For contemporary agriculturalists, the range is between 1 (Rosetta 1988) and 27% (De Garine & Koppert 1988). Estimates of intakes of protein, iron and calcium based on diets in which animal foods supply different proportions of the energy intake in the two periods are given in table 5.

When the diet is adequate to sustain energy balance at a BMI of 20.5 kg m^{-2}, intakes of protein are high, for both model men. If the proportion of dietary energy from animal sources was 80% in pre-Neolithic times, the proportion of dietary energy from protein would have been very close to 50%, a level at which toxicity from excessive ingestion of nitrogenous compounds could ensue (Speth, this symposium). Lower levels of meat consumption would have resulted in considerably lower protein:energy ratios than this ceiling value.

An iron-rich intestinal environment may predispose human hosts to infection (Stockman 1981), and it is possible that human populations in the pre-Neolithic Near East and Mediterranean may have suffered from high levels of infection and intestinal infestation partly as a result of high intakes of iron. This could have been a contributing factor to certain disease pathologies observed in skeletal material from this region, and one which has not been considered by palaeoanthropologists. Iron intakes in the Neolithic model man are above the 5–9 mg per day recommended by FAO/WHO (1974) as the minimum to prevent anaemia in a population. The only possible way in which the anaemias reported by Angel (1984) could have been due directly to nutritional deficiencies of iron would have been if the intake of animal foods was nil, or very low indeed. Under such conditions, the intake of dietary iron, although superficially adequate, is overwhelmingly non-haem, and thus poorly absorbed. It is more likely that the pathologies described by various authors were due to anaemias which arose out of intestinal parasitism and other infectious agents and not nutritional shortfall, at least under conditions where subjects were not chronically energy deficient.

Calcium intakes of both model men are lower than the FAO/WHO (1974) recommended minimum values of 400–500 mg d^{-1} at all levels of meat consumption. Calcium balance can be maintained at a variety of levels of calcium intake, however. The highest figure recorded in the literature is 975 mg d^{-1} (Heaney *et al.* 1978), whereas the lowest is 200 mg d^{-1}, in a sample of 10 Peruvian men (Hegsted *et al.* 1952). Although there is enormous variation in absorptive efficiency of calcium, the smallest amount that needs to be absorbed by an adult male to stay in calcium homeostasis is about 100 mg d^{-1} (Stini 1990). This is approximately the amount that is lost daily in urine.

It is possible that the values for calcium intake reported here could be adequate to maintain calcium balance in both periods, if populations were extremely efficient in their absorption of this mineral. There are two reasons why this may not be the case, however. The first is that bioavailability of calcium is lower on high protein diets than on lower protein diets (Yuen *et al.* 1984, Kersetter & Lindsay 1990). This condition could have prevailed in pre-Neolithic times, if levels of meat consumption were high. The second is that the calcium in plant foods is not particularly well absorbed, due to the presence of phytate and oxalate, among other substances (McCance & Widdowson 1942; Cummings *et al.* 1979; Fincke & Sherman 1935). This condition is likely to have prevailed in the Neolithic, if the intake of animal foods was particularly low.

It is possible that the osteoporosis observed in the skeletal record from the Neolithic period could have been due to vitamin D deficiency or phosphorus excess, as well as primary calcium deficiency. Dietary intakes of vitamin D were probably low in both pre-Neolithic and Neolithic periods, as dairy foods played no part in the diet. Phosphorus intakes were more likely to be higher in pre-Neolithic times, if large amounts of meat were eaten. Therefore it might be expected that pre-Neolithic man would have suffered from osteoporosis in equal or greater measure than Neolithic man, if the condition was due primarily to dietary factors. The palaeopathological record does not support this assertion.

5. THE EFFECTS OF ENERGY UNDERNUTRITION ON THE AVAILABILITY OF DIETARY IRON AND CALCIUM

To examine the possible effects of CED on the availability of iron and calcium in the diet of subjects living in the Neolithic, the model was re-run under modified conditions. The following assumptions were made: (i) that the model man suffers from CED which is manifested in low BMI and possibly low PAL; (ii) that the types and relative proportions of foods eaten are the same as those in the previous run, at all levels of BMI; (iii) that in the estimation of PAL, there is no down-regulation of BMR due to CED; and (iv) that subjects are in energy balance at the lower levels of intake, where energy intake is equal to TEE. Results are given in table 6.

At the higher PAL ($2.0 \times$ BMR), levels of protein and iron intake are adequate to high, regardless of BMI or the proportion of the diet coming from animal foods. At low PAL ($1.35 \times$ BMR), protein intakes are lower but still adequate, whereas iron intakes are within the range of values recommended by FAO/WHO (1974) in the 0 and 5% animal foods categories, and above this range for the 10% animal foods group. It is possible, however, that at very low levels of animal food intake, the iron in the diet would have been poorly absorbed. Calcium intakes are consistently low, with values falling below 200 mg d^{-1} at PAL of $1.35 \times$ BMR. This may be a level below which calcium balance cannot be sustained; if so, the osteoporosis observed in the Neolithic skeletal record can only be attributed to primary calcium deficiency under conditions of moderate or severe CED.

6. DISCUSSION

The transition from hunting and gathering to cultivation and animal domestication in the Near East and Mediterranean involved greater utilization of energy-rich plant foods to the detrement of energy-poor ones, and an enormous reduction in the consumption of animal foods. This dietary change has been associated with overall dietary deficiency, and specific deficiencies of protein (Angel 1984), calcium (Smith *et al.* 1984), and possibly iron (Rathbun 1984) by inference from the examination of skeletal material from pre-Neolithic and Neolithic times.

The modelling of dietary and nutritional intakes of an adult male in pre-Neolithic times and in the Neolithic has allowed the postulated association between dietary change and nutrient deficiencies in the Neolithic to be tested. The results suggest that protein deficiency was unlikely if CED was not being experienced; that is, if mass in relation to height was not low, even if adult males were short statured. If adult males suffered from CED, then dietary protein deficiency was only likely if the diet contained no meat. Deficiency of dietary calcium was possible in the Neolithic, if populations were experiencing moderate or severe CED. In the absence of CED, it is unlikely that dietary iron deficiency was a problem. Even if adult males in the Neolithic were suffering CED, it is unlikely that the anaemias identified by palaeoanthropologists had a specific dietary aetiology, although a more complex dietary aetiology is possible.

I thank Dr J. Boldsen, Dr C. J. K. Henry, Professor M. Jones, Ms M. Lahr and Dr S. S. Strickland for their helpful comments and criticisms of an earlier draft of this paper.

REFERENCES

Angel, J. L. 1971 *The People of Lerna.* Washington D.C.: Smithsonian Institution.

Angel, J. L. 1984 Health as a crucial factor in the changes from hunting to developed farming in the Eastern Mediterranean. In *Palaeopathology at the origins of agriculture* (ed. M. N. Cohen & G. J. Armelagos), pp. 51–73. New York: Academic Press.

Armelagos, G. 1987 Biocultural aspects of food choice. In *Food and evolution. Toward a theory of human food habits* (ed. M. Harris & E. B. Ross), pp. 579–594. Philadelphia: Temple University Press.

Bar-Yosef, O. 1987 Late Pleistocene adaptations in the Levant. In *The Pleistocene Old World: regional perspectives* (ed. O. Soffer), pp. 219–236. New York: Plenum Press.

Bokonyi, S. 1969 Archaeological problems and methods of recognising animal domestication. In *The Domestication and exploitation of plants and animals* (ed. P. J. Ucko and G. W. Dimbleby), pp. 219–229. London: Duckworth.

British Nutrition Foundation Task Force on Calcium. 1989 London: British Nutrition Foundation.

Brun, T., Bleiberg, F. & Goihman, S. 1981 Energy expenditure of male farmers in dry and rainy seasons in Upper-Volta. *Br. J. Nutr.* **45**, 67–75.

Brun, T. A., Geissler, C. A., Mirbagheri, I., Hormozdiary, H., Bastani, J. & Heydayat, H. 1979 The energy expenditure of Iranian agricultural workers. *Am. J. clin. Nutr.* **32**, 2154–2161.

Butzer, K. W. 1971. *Environment and archaeology: an introduction to Pleistocene geography.* Chicago: Aldine.

Cleave, J. H. 1970 *Labour in the development of African agriculture: the evidence of farm surveys.* Ph.D. thesis, Stanford University.

Clutton-Brock, J. 1989 *The walking larder. Patterns of domestication, pastoralism, and predation.* London: Unwin Hyman.

Cohen, M. N. 1989 *Health and the rise of civilisation.* New Haven: Yale University Press.

Cummings, J. H., Hill, M. J., Houston, M. N., Branch, W. J. & Jenkins, D. J. A. 1979 The effect of meat protein and dietary fiber on colonic function and metabolism. 1. Changes in bowel habit, bile acid excretion and calcium absorption. *Am. J. clin. Nutr.* **32**, 2086–2093.

Dasgupta, B. 1977 *Village society and labour use.* Delhi: Oxford University Press.

De Garine, I. & Koppert, G. 1988 Coping with seasonal fluctuations in food supply among savanna populations: the Massa and Mussey of Chad and Cameroon. In *Coping with uncertainty in food supply* (ed. I. de Garine & G. A. Harrison), pp. 210–259. Oxford University Press.

FAO 1982 *Food composition tables for the Near East* (FAO Food and Nutrition Paper No. 26). Rome: Food and Agriculture Organisation.

FAO/WHO 1974 *Handbook on human nutritional requirements.* Geneva: World Health Organisation.

FAO/WHO/UNU 1985 *Energy and protein requirements* (Technical Report Series 724). Geneva: World Health Organisation.

Fincke, M. L. & Sherman, H. C. 1935 The availability of calcium from some typical foods. *J. biol. Chem.* **110**, 421–428.

Flannery, K. V. 1965 The ecology of early food production in Mesopotamia. *Science, Wash.* **147**, 1247–1256.

Flannery, K. V. 1969 Origins and ecological effects of early domestication in Iran and the Near East. In *The domestication and exploitation of plants and animals* (ed. P. J. Ucko & G. W. Dimbleby), pp. 73–100. London: Gerald Duckworth & Co. Ltd.

Fox, R. H. 1953 *A study of the energy expenditure of Africans engaged in various activities, with special reference to some environmental and physiological factors which may influence the efficiency of their work.* Ph.D. thesis; University of London.

Guet, G. 1960 *Dario: village de Haute-Sangha.* Paris: Bureau d'Etudes et de Recherches du Plan.

de Guzman, P. E., Dominguez, S. R., Kalaw, J. M., Basconcillo, R. O. & Santos, V. F. 1974 A study of the energy expenditure, dietary intake, and pattern of daily activity among various occupation groups. I. Laguna rice farmers. *Philippine J. Sci.* **103**, 53–65.

Harris, D. R. & Hillman, G. C. 1989 *Foraging and farming. The evolution of plant exploitation.* London: Unwin Hyman.

Heaney, R. P., Recker, R. R. & Saville, P. D. 1978 Menopausal changes in calcium balance performance. *J. Lab. clin. Med.*, **92**, 953–963.

Hegsted, D. M., Muscoso, I. & Collazos, C. 1952 Study of minimum calcium requirements of adult men. *J. Nutr.* **46**, 181–201.

Hill, K. 1982 Hunting and human evolution. *J. hum. Evol.* **11**, 521–544.

Hill, K., Kaplan, H., Hawkes, K. & Hurtado, A. M. 1985 Men's time allocation to subsistence work among the Ache of Eastern Paraguay. *Hum. Ecol.* **13**, 1, 29–47.

Hillman, G. C., Colledge, S. M. & Harris, D. R. 1989 Plant food economy during the Epipalaeolithic period at Tell Abu Hureyra, Syria: dietary diversity, seasonality, and modes of exploitation. In *Foraging and farming. The evolution of plant exploitation* (ed. D. R. Harris & G. C. Hillman), pp. 240–268. London: Unwin Hyman.

Hipsley, E. H. & Kirk, N. E. 1962 *Studies of dietary intake and the expenditure of energy by New Guineans* (South Pacific Commision Technical Paper No. 147). Noumea.

Hopf, M. 1969 Plant remains and early farming in Jericho. In *The domestication and exploitation of plants and animals* (ed. P. J. Ucko & G. W. Dimbleby), pp. 355–359. London: Duckworth.

Huss-Ashmore, R. A., Goodman, A. H. & Armelagos, G. J. 1982 Nutritional inference from paleopathology. *Adv. Archaeol. Meth. Theory* **5**, 395–474.
Jacobsen, T. W. 1969 Excavations at Porto Cheli and vicinity, preliminary report. II. The Franchthi Cave, 1967–1968. *Hesperia* **38**, 343–381.
James, W. P. T., Ferro-Luzzi, A. & Waterlow, J. C. 1988 Definition of chronic energy deficiency in adults. *Eur. J. clin. Nutr.* **42**, 969–981.
Kersetter, J. E. & Lindsay, H. A. 1990 Dietary protein increases urinary calcium. *J. Nutr.* **120**, 134–136.
Kurth, G. & Rohrer-Ertl, O. 1981 On the anthropology of the Mesolithic to Chalcolithic human remains from the Tell es-Sultan in Jericho, Jordan. In *Excavations at Jericho*, Vol. III (Ed. K. M. Kenyon), pp. 407–499. Jerusalem: British School of Archaeology.
McCance, R. A. & Widdowson, E. M. 1942 Mineral metabolism of healthy adults on white and brown bread dietaries. *J. Physiol.* **101**, 44–85.
McNeill, G., Payne, P. R. & Rivers, J. P. W. 1987 *Patterns of adult energy nutrition in a South Indian village.* (Department of Human Nutrition Occasional Paper No. 11.) London School of Hygiene and Tropical Medicine.
Montgomery, E. & Johnson, A. 1974 Machiguenga energy expenditure. *Ecol. Food Nutr.* **6**, 97–105.
Moore, A. T. 1985 The development of Neolithic societies in the Near East. *Adv. World Archaeol.* **4**, 1–70.
Moseley, J. E. 1965 The paleopathologic riddle of symmetrical osteoporosis. *Am. J. Roentgenology* **95**, 135–143.
Nathan, H. & Haas, N. 1966 Cribra orbitalia. A bone condition of the orbit of unknown nature. *Israel J. Med. Sc.* **2**, 171–191.
Norgan, N. G., Ferro-Luzzi, A. & Durnin, J. V. G. A. 1974 The energy and nutrient intake and the energy expenditure of 204 New Guinean adults. *Phil. Trans. R. Soc. Lond.* B **268**, 309–348.
Paul, A. A. & Southgate, D. A. T. 1978 *McCance and Widdowson's The Composition of Foods.* London: HMSO.
Rathbun, T. A. 1984 Skeletal pathology from the Paleolithic through the metal ages in Iran and Iraq. In *Paleopathology at the origins of agriculture* (ed. M. N. Cohen & G. J. Armelagos), pp. 137–167. New York: Academic Press.
Reed, C. A. 1977 A model for the origin of agriculture in the Near East. In *Origins of agriculture* (ed. C. A. Reed), pp. 543–567. The Hague: Mouton.
Rosetta, L. 1988 Seasonal variations in food consumption by Serere families in Senegal. *Ecol. Food Nutr.* **20**, 275–286.
Schofield, W. N. 1985 Predicting basal metabolic rate, new standards and review of previous work. *Hum. Nutr. Clin. Nutr.* **39**C (Suppl. 1), 5–41.
Smith, P., Bar-Yosef, O. & Sillen, A. 1984 Archaeological and skeletal evidence for dietary change during the late Pleistocene/early Holocene in the Levant. In *Paleopathology at the origins of agriculture* (ed. M. N. Cohen & G. J. Armelagos), pp. 101–136. New York: Academic Press.
Stini, W. A. 1990 'Osteoporosis': etiologies, prevention, and treatment. *Yb. phys. Anthropol.* **33**, 151–194.
Stockman, J. A. 1981 Infections and iron. *Am. J. Dis. Child.* **135**, 18–20.
Stuart-Macadam, P. L. 1991 Anemia in past human populations. *Am. J. phys. Anthropol.* **85**, 170–171.
Suzuki, S. 1988 Villagers' daily life and the environment. In *Health ecology in Indonesia* (ed. S. Suzuki), pp. 13–22. Tokyo: Gyosei Corporation.
Tin-May-Than & Ba-Aye 1985 Energy intake and energy output of Burmese farmers at different seasons. *Hum. Nutr. Clin. Nutr.* **39**C, 7–15.
Trotter, M. & Gleser, G. C. 1952 Estimation of stature from long bones of American whites and negroes. *Am. J. phys. Anthropol.* **10**, 463–514.
Ulijaszek, S. J. 1987 *Nutrition and anthropometry – with special reference to populations in Papua New Guinea and the United Kingdom.* Ph.D. thesis, University of London.
Van Zeist, W. & Bakker-Heeres, J. A. H. 1979 Some economic and ecological aspects of the plant husbandry of Tell Aswad. *Paleoorient* **5**, 161–169.
Viteri, F. E., Torun, B., Garcia, J. C., & Herrera, E. 1971 Determining energy costs of agricultural activities by respirometer and energy balance techniques. *Am. J. clin. Nutr.* **24**, 1418–1430.
Waterlow, J. C. 1986 Notes on the new estimates of energy requirements. *Proc. Nutr. Soc.* **45**, 351–360.
Wing, E. S. & Brown, A. B. 1979 *Paleonutrition. Method and theory in prehistoric foodways.* New York: Academic Press.
Yuen, D. E., Draper, H. H. & Trilok, G. 1984 Effect of dietary protein on calcium metabolism in man. *Nutr. Abs. Rev.* **54**, 447–459.

Discussion

G. Hillman (*Department of Human Environment, Institute of Archaeology, University College London, U.K.*). At the start of his paper, Dr Ulijaszek made reference to the possible role of phytates in limiting the uptake of calcium and thereby contributing to osteoporosis and other pathologies. However, he eventually concluded that dietary factors are unlikely to have contributed to these pathologies.

I am not a nutritionist and cannot comment on any links between calcium malabsorbtion and osteoporosis. However, I am familiar with the existing evidence for changes in the plant-based components of diet which seem to have accompanied the shift from foraging to farming in Dr Ulijaszek's study area (Southwest Asia). My present reading of this evidence strongly suggests that the shift to cultivation would have involved a dramatic increase in the consumption of phytate-rich foods (probably as year-round staples), and presumably, therefore, a correspondingly increased risk of chronic calcium malabsorption.

S. J. Ulijaszek. Phytate intakes were not incorporated into the model, because the phytate content of only a limited range of foods has been thus far estimated. Certainly, higher intakes of phytate in the Neolithic, in association with marginal intakes of calcium, may have played an important part in the aetiology of osteoporosis at this time.

J. L. Boldsen (*Department of Community Health, University of Odense, Denmark*). I think that the height of the Neolithic people has been overestimated by using mainstream regression formulae (Trotter & Gleser 1952). Would Professor Ulijaszek comment on the effect on calcium and iron balance of the human height of the neolithic populations being 165 cm instead of 170 cm?

S. J. Ulijaszek. At a smaller body size, the Neolithic model man would maintain energy balance at lower levels of food intake, and this would result in lower levels of intake of both iron and calcium. It is unlikely that iron intakes would be low enough to cause iron deficiency anaemia, unless the model man were suffering chronic energy deficiency. However, the possibility of primary calcium deficiency would be greater than that suggested in the presentation.

C. J. Henry (*School of Biology, Oxford Polytechnic, U.K.*). Much of Dr. Ulijaszek's discussion has centred around calcium intake in early man. What relevance does the lack of vitamin D have on calcium metabolism in the Neolithic man? Could vitamin D have also been a limiting micronutrient thus influencing calcium metabolism in this population?

S. J. ULIJASZEK. Certainly, dietary intakes of vitamin D would have been low in the Neolithic, because dairy products played little or no part in the diet. However, this vitamin can be synthesized in the skin from 7-dehydrocholesterol by the action of ultraviolet light. Climatological evidence indicates that overall rainfall in the Near East and Mediterranean region was lower in the Neolithic than in present times, suggesting a drier, sunnier clime. It is probable that on average, human groups received plentiful exposure to sunlight in the course of work, making the possibility that vitamin D deficiency was a major contributing factor to osteoporosis unlikely.

Nature and variability of human food consumption

D. A. T. SOUTHGATE

AFRC Institute of Food Research, Norwich Laboratory, Norwich Research Park, Colney, Norwich NR4 7UA, U.K.

SUMMARY

The early human diet was characteristically extremely varied, and a wide range of plant species and plant organs were consumed. Foods of animal origin included those taken opportunistically, such as invertebrates, amphibians, reptiles, small mammals, birds and their eggs, and the scavenging and hunting of larger mammals. Each of these types of food have characteristic nutritional compositions. Comparison of these compositional features shows that an adequate diet could be obtained in many different ways. The selection of food providing fat had substantial advantages in reducing the amount of plant foods to be gathered, in the satiety provided and in supplying essential micronutrients. Obtaining adequate water and energy would probably be the main physiological drives. Many plant foods contain natural toxicants, and would only have been suitable as major items in the diet once cooking had been developed, and the preference for sweet tastes would have protected humans from eating bitter, toxic plants.

1. INTRODUCTION

The human diet is characteristically extremely varied, and the ability to select a dietary mixture from a wide range of edible plants and animals, coupled with great ingenuity in making intrinsically unpromising materials into nutritious, if not necessarily palatable foods, can be seen as an important factor in the development of human societies (Le Gros Clark 1967). The capacity to identify and to make edible, natural materials in the most hostile of environments has enabled the human species to expand its range. The acceptance of variety was also a key factor in surviving in times of extreme food shortages that were common before the development of the continuity of food supply, which characterizes the present position in developed industrial societies.

In discussing the 'natural' diet of humans I think that this characteristic of extreme fluctuations in the amounts and types of foods available for consumption is an important one, and that the control of appetite and the choice of foods evolved under conditions of alternating abundance and scarcity, circumstances which are still the norm in many developing countries today.

In this paper I shall use 'natural' for diets composed of plant foods that have not been cultivated in any developed agricultural system and foods derived from undomesticated animals (Southgate 1988). It would be incorrect to think of these foods as unprocessed because, as I will show, the development of the human diet has been characterized by the use of processing technologies, albeit of a simple nature by modern standards, but of great dietary significance when they were introduced.

I shall first discuss the range and compositional features of the various major types of food that were available to early human societies. In this I will draw on the many studies that have been made on surviving populations of hunter–gatherers. In doing this I recognize that these societies have undoubtedly developed themselves, and that they have had extensive contacts with agricultural and pastoral societies. It would be naïve to assume that these contacts have not affected their food patterns, and it would be equally naïve to believe that their societies and environment have been stable (Schrire 1984). It is important to recognize that the residual populations of hunter–gatherers have to a great extent been driven, by the development of agriculture and the consequential reduction in their hunting ranges, into less productive environments. Second, I shall discuss briefly the factors influencing the choice of the foods. I shall then discuss how the composition of the various foods, when considered against the nutritional needs of humans, imposes constraints on dietary selection.

This analysis raises some hypotheses regarding the evolution of the determinants of food choice.

2. THE FOODS AVAILABLE: THEIR COMPOSITION AND OTHER ATTRIBUTES

It is reasonable to assume that the types of foods eaten by modern higher primates provide a starting point for considering the foods eaten by early human populations, although one must also recognize that primate societies have also developed and their eating patterns may have developed alongside other changes (Jolly 1985). These, taken with the range of foods eaten by present-day hunter–gatherers, provide the basis for this analysis.

(a) *Plant foods*

A wide range of plant foods is consumed including most parts of the plant, such as fruits, seeds, leaves, roots and tubers. Studies of many present hunter–gatherers show that the numbers of plant species

Table 1. *Compositional features of plant foods/grams per 100 g edible matter*

	fruits	nuts	seeds	seed legumes	leafy vegetables	roots and tubers
water	61.0–89.1	4.5–51.7	12–15	74.6–80.3	84.3–94.7	62.3–94.6
protein	0.3–1.1	2.0–24.3	8.4–13.6	5.7–6.9	0.2–3.9	0.1–4.9
fat	trace–4.4	2.7–64.0	1.5–5.4	1.0–15	0.2–1.4	0.1–0.4
sugar	4.4–34.8	3.2–7.0	traces	1.8–3.2	1.5–4.9	0.5–9.5
starch	trace–3.0	1.8–29.6	72.9–73.9	5.4–8.1	0.1–0.8	11.8–31.4
dietary fibre (plant cell wall)	2.0–14.8	5.7–14.3	4.0–12.0	4.5–4.7	1.2–4.0	1.1–9.5
energy/kJ	90–646	720–2545	1356–1574	247–348	65–177	297–525
energy/kcal	22–155	170–639	324–376	59–83	24–43	71–138
micronutrients	vitamin C carotenoids K, Mg	B-vitamins vitamin E K, Mg, P, Fe	B-vitamins vitamin E K, Mg, P, Fe	B-vitamins vitamin C K, Mg, P, Fe	vitamin C, folates carotenoids Ca, Fe	vitamin E carotenoids Fe, K, Ca
toxic constituents	cyanogenetic glycosides in seeds	—	—	haemoagglutonins, anti, tryptic, lectins	glucosinolates	glycoalkaloids

Sources: Paul & Southgate (1978); Brand *et al.* (1983); Brand *et al.* (1985); Kuhnlein, (1989; 1990); Woodburn & Southgate (unpublished data).

collected is high: Gonzalez (1972), in studies of North American Indians, found that more than 400 species were collected, and of these 130 were consumed as foods. The remainder were collected for their herbal properties in relation to medicinal or ritual use. The nutritional compositions of the different plant organs are distinctive and need to be considered separately.

(*i*) *Fruits*. Some of the key compositional features of a range of fruits are given in table 1. Characteristically fruits have a high water content, and contain low levels of protein and fat; the protein, moreover, is often concentrated within the seeds, which are usually lignified and therefore resistant to both digestion in the small intestine and bacterial degradation in the large intestine, and is therefore unavailable to the body. The major nutrients provided by fruits are carbohydrates; fruits characteristically contain fructose, glucose and sucrose, but usually little or no starch when mature. The other carbohydrates present are derived from the plant cell walls, and these are usually parenchymatous, relatively undifferentiated structures (Selvendran 1984). As such they are rich in soluble polysaccharides (pectic substances) and are readily degraded by the microflora of the large intestine (Cummings *et al.* 1979). Fruits, therefore, contribute metabolizable energy as carbohydrate from the sugars, and small amounts of energy as free fatty acids from the cell wall materials degraded in the large bowel. They are also important sources of essential micronutrients, especially vitamin C and carotenoids (provitamin A).

Fruits, as foodstuffs, are available for a limited time, and when ripe, and most acceptable, are sometimes difficult to collect and transport. Many fruits, therefore, are best consumed at the time and place they are picked, whereas others lend themselves to being gathered in the strict sense. When ripe they have a short period of acceptability before senescence intervenes.

(*ii*) *Nuts and seeds*. Nuts and seeds, especially the former, are important in the diet of primates and present hunter–gatherers (Brand *et al.* 1985). Their compositional features are given in table 1. These foods are characterized by having thickened, usually lignified, outer seed coats. Many seeds have less rigid testae that contain high levels of polyphenolic materials such as tannins. These protect the seed from desiccation and, more importantly, from fungal and insect damage: they also protect the seed from bacterial degradation in the rumen and large bowel of animals consuming them. Compositionally seeds are rich in protein in the embryo, and contain a store of either lipids or polysaccharides (which may include starch) which serves as an energy source for the germinating plant. The plant cell wall material of seeds includes thin-walled structures in the endosperm or cotyledons and thicker lignified outer coats. Seeds form the most energy- and protein-rich plant foods.

Seeds have much lower water content than all other plant foods, and are therefore stable and easily gathered, transported and stored.

The most important seeds consumed in the present human diet are those of the cereals. It is possible that grains of wild cereals were part of the diet of early humans but it is reasonable to assume (Eaton & Konner 1985) that cereals only began to be significant in the human diet within the last 9000 years with the development of agriculture.

The seeds of the Leguminosea form a distinct group compositionally; they may be eaten as mature seeds, sharing many of the characteristics of other seeds: that is, being rich in protein and either lipids or polysaccharides with a high nutrient density. They may also be consumed immature with the seed pod when the nutrient densities are correspondingly lower. However, many legumes contain toxic plant metabolites, haemoglutenins, saponins and lectins, and some of them are very toxic indeed (Liener 1969; Silverstone 1985). The wider use of these very nutritious foods would only have become possible when foods began to be cooked in water, so that they would only have become significant in the human diet relatively late (Stahl 1984). To use the nuts and seeds as foodstuffs, the outer shells or coats must be broken and this would have marked the first type of food processing. Without this treatment the nutrients would not become available. Additionally the seed husks that are rich in polyphenolic materials impart a bitter taste to the broken seed, so some type of winnowing would be required. Mature cereal grains from wild cereals are virtually indigestible unless the grain is cracked or soaked in water, and maximum nutritional value requires reduction of particle size by grinding; the value of cereals would therefore be limited until simple querns were developed.

(*iii*) *Leaves and stems*. These are widely consumed, especially the emerging buds, by many primates, and leafy plants are widely consumed by humans. Table 1 summarizes the main compositional features of these structures. The protein contents are higher than fruits, and they contain lower amounts of sugars. Some leaves contain small amounts of starch but the plant cell walls are the major polysaccharides present; these are similar to those found in fruits. Leaves, however, are often characterized by having cutinized epidermal layers which restrict the degradation of the polysaccharides by the intestinal flora. The effective extraction of the soluble contents of the leaves and stems also requires extensive mechanical disruption of the cellular structures (Pirie 1973) which would not be achievable with human dentition. The maximal nutritional use of leaves as a source of protein requires either the assistance of a microflora such as that present in the rumen of ruminant animals or the stomachs of the leaf-eating monkeys. The range of micronutrients present include vitamin C, folates and carotenoids, as well as iron and calcium.

Young leaves lend themselves to consumption at the time of gathering. Both leaves and stems are bulky material to transport and are not very stable when stored.

Many plants produce secondary metabolites that have bitter or astringent properties, and many produce toxic alkaloidal and other compounds such as haemoglutenins. Others produce intestinal enzyme inhibitors and compounds, such as lectins, which bind to the mucosal surfaces and inhibit digestion, especially of proteins (Leiner 1969). These natural toxicants would, firstly, make many plants unattractive as foods from a sensory point of view and, secondly, often produce profound intestinal discomfort when eaten. Present day hunter–gatherers recognize these unattractive and toxic foods and may collect them for use as herbal medicines or as sources of poisons for use in hunting or fishing. Until fire was discovered and foods were heated, a process which inactivates or destroys many of these components, the use of many leaves would have been limited (Stahl 1984).

(*iv*) *Roots and tubers*. These are important in the diet of many present hunter–gatherers. The composition of roots and tubers is given in table 1. Many tubers and some roots contain significant amounts of starch. Roots tend to contain more sugar than starch, and the major sugar is often sucrose. The plant cell walls are undifferentiated and rich in soluble polysaccharides. Vitamin levels are lower than in the aerial parts of plants. Some roots such as cassava (manioc) contain toxic secondary metabolites and require soaking in water before they are safe to consume. As foodstuffs roots, and especially tubers, can be time-consuming to collect but they are often protected from desiccation and fungal damage by external suberinized layers, and so can be stored for long periods.

(*v*) *Fungi*. Some fungi are highly valued as foods by present hunter–gatherers; many, however, are extremely toxic, and societies that consume fungi have learned to avoid the more toxic varieties or to restrict their use to medicinal or ritual purposes.

(*b*) *Foods of animal origin*

By analogy with the behaviour of non-human primates it is probable that foods of animal origin were widely consumed. The activities involved in collecting plant foods would lead to the opportunist taking of a wide range of animals; insects and their grubs, for example, are eaten by many communities. Honey is a highly prized food in many hunter–gatherer societies. It provides the sweetest food that is available to hunter–gatherers; it is a seasonal crop and the harvests are associated with festivals. The yields from a nest range from 1 to 30 kg, and the honey frequently contains dead bees and grubs (E. V. Crane, personal communication).

Eggs and fledgling birds would also be found during the collection of plant foods, as would reptiles such as lizards, although the consumption of snakes appears less likely. Communities close to water would also find a wide range of molluscs and Crustacea which could be harvested readily. The shell middens associated with many human sites show that molluscs were extensively consumed (Schrire 1984). These animal foods could be obtained without the need for weapons or tools, and may have been eaten very early in human evolution. One can also speculate that the young of many small animals, for example rodents, would be found during searching for tubers, and these can be skinned using the hands alone, as can young birds.

Large animals would have required the development of organized hunting, initially using trapping methods with pits or by driving animals towards cliffs or ravines. Tools for skinning these animals and for butchering carcasses into suitable joints for transport would have been a pre-requisite for these foods being available. The development of weapons for killing at a distance would have made hunting less hazardous and more effective. Chimpanzees exhibit effective organized hunting skills involving driving animals towards flanking and stopping members of the troop, and it is reasonable to assume that similar skills were developed early in human evolution.

Communities near water developed analogous skills in catching fish and other aquatic animals. In considering the composition of these foods it is convenient to consider them in four major categories: invertebrates; amphibians, reptiles, eggs and birds; fish; and mammals.

(*i*) *Invertebrates*. These include insects, both mature and as grubs, Crustacea and molluscs (table 2). These

Table 2. *Composition of invertebrates/grams per 100 g*

	insects			
	grubs	mature	molluscs[a]	Crustacea[a]
water	36.6–66.6	35.2–62.2	75.8–84.6	76.8–82.0
protein	8.7–18.6	20.5–26.8	8.4–18.7	14.6–18.1
fat	13.2–37.5	3.8–38.8	0.2–2.2	0.5–2.0
carbohydrates[b]	6.9–16.0	1.4–5.5	traces	traces
energy/kJ	833–1760	715–1912	289–410	285–418
energy/kcal	199–420	170–457	69–98	69–100
micronutrients	B-vitamins K, Mg, Ca, Fe, Zn	B-vitamins Fe, Zn	B-vitamins (B_{12}) vitamin E Fe, Zn	B-vitamins Ca, Mg, P, Fe, Zn

[a] Edible matter only.
[b] These are principally chitins from exoskeleton.
Source: Cherikoff *et al.* (1985); Wu Leung (1968); Paul & Southgate (1978).

Table 3. *Composition of amphibians, birds and eggs/grams per 100 g edible matter*

	frog[a]	lizard[a]	snake[a]	turtle[a]	young birds	eggs[b]
water	83.6	58.8–73.2	75.0	71.7–80.9	66.2–68.2	70.4–79.8
protein	15.3	24.6–31.6	14.4	17.5–25.4	20.2–23.7	10.7–13.9
fat	0.3	1.2–11.3	3.3	0.2–1.5	8.5–9.6	7.0–14.2
carbohydrate	trace	trace	trace	traces	traces	traces
energy kJ	285	472–906	393	343–469	742–749	485–787
energy kcal	68	113–217	94	82–112	175–179	116–188
micronutrients	B-vitamins P, Fe, Zn	B-vitamins P, Fe, Zn	B-vitamins P, Fe, Zn	B-vitamins P, Fe, Zn	B-vitamins (B_{12}) Fe, Zn	B-vitamins retinol Ca, Fe, Zn

[a] These animals have fat depots as isolated organs which contain 40–60% fat.
[b] Reptile eggs are similar in composition.
Source: Cherikoff *et al.* (1985); Wu Leung (1968).

Table 4. *Composition of some mammalian sources/g per 100 g edible meat*

	rabbit	squirrel	pig[a]	horse	buffalo	heart	kidney	liver
water	74.3	72.2	—	70.0	76.5	73.6–79.2	78.8–79.8	67.3–72.9
protein	16.9	26.3	10–14.0	19.0	17.7	15.2–17.1	15.7–17.1	19.1–21.3
fat	7.9[b]	0.4[b]	35.0–55.0	10.0[b]	4.9[b]	2.6–9.3	2.6–2.7	6.3–10.3
carbohydrate	—	trace	trace	glycogen	—	—	—	glycogen in fresh
energy kJ	594	460	1550–2240	713	502	363–629	363–380	567–683
energy kcal	142	110	371–535	170	120	86–150	86–150	153–179
micronutrients	B vitamins B_{12} retinol Ca, Fe, Zn	B vitamins B_{12} retinol Ca, Fe, Zn	B vitamins B_{12} retinol Ca, Fe, Zn	B vitamins B_{12} retinol Ca, Fe, Zn	B vitamins B_{12} retinol Ca, Fe, Zn	B vitamins B_{12} retinol Ca, Fe, Zn	B vitamins B_{12} retinol Ca, Fe, Zn	B vitamins B_{12} retinol Ca, Fe, Zn

[a] Whole carcass.
[b] Fat associated with muscle. Fat deposits would also be present.

foods are characteristically rich sources of protein, although they have a high moisture content; some insect grubs are rich sources of fat but in the other invertebrates fat contents are low. They are relatively small animals, and the insects were probably eaten when taken; the other invertebrates would be reasonably easy to collect.

(*ii*) *Amphibians, reptiles, eggs and birds* (see table 3). The smaller amphibians and reptiles share many of the compositional characteristics of the invertebrates: high water content, with modest protein and low fat levels. Eggs, however, are relatively rich sources of both protein and fats. Eggs of all species are good sources of micronutrients, especially the fat-soluble vitamin A (retinol). Young birds, especially just before they start to fly, have substantial fat stores, and many present-day communities with access to large nesting colonies harvest the young birds and render them to produce fats (Chambers & Southgate 1969).

(*iii*) *Mammals*. The carcass of an animal provides a wide range of types of food. In bulk terms the musculature is the major part, providing meat and some fat. Most wild animals have a lower total carcass fat than domesticated species, and the fats tend to be less saturated even in the ruminant animals (Crawford 1968; Talbot 1964). There is less 'marbling' of fats within the muscles, fat storage being subcutaneous, mesenteric and perirenal. Organs, especially liver, kidney and heart, are rich sources of protein and many micronutrients. Bones, in addition to providing material for tools and ornaments, contain in their marrow a rich source of protein, fat and iron. Typical values are given in table 4.

(*iv*) *Fish*. The fishes forming part of the diet of early man would have been primarily freshwater or estuarine species, and may have included both fatty and non-fatty white fish. Fish are characteristically rich sources of protein, and the fatty fish, such as salmonoids returning to fresh waters to spawn, are rich in fats and the fat-soluble vitamins.

Before concluding this section it is important to recognize one major animal food that would have been part of the natural diet of virtually all humans, that is breast milk. The milk of all species is highly characteristic, and human milk is no exception. It has a low protein content, and fat provides around 50% of the total energy. All milks contain the sugar lactose, and it is almost certain that this sugar would only have been present in the diet of early humans during the suckling period, which may have been in excess of two years.

3. THE DETERMINANTS OF FOOD CHOICE

It is important to consider briefly some of the factors that would have acted as the major determinants of food choice from the range of foods in the natural environment.

(*a*) *Availability*

This would have been a very important factor; even in the tropical and sub-tropical regions many foods are only available for consumption at certain times during the year. The gathering of sufficient amounts of plant foods involves travelling over considerable distances, and so the size of the range and its productivity determine food availability. The pattern of consumption of plant and animal foods in some existing hunter–gatherers is very closely determined (Coles-Rutishauser 1986) and seasonal variations in the types and amounts of foods consumed must have been a common feature of life.

(*b*) *Sensory preferences*

The primary senses have a major role in the selection of foods, and the use of visual appearance is a key factor in identifying fruits at the desired stage of maturity and selecting the youngest and most tender shoots. The learned associations between appearances and unpleasant tastes or after-effects of eating plant foods or insects and other animals are very important. Thus the preference for sweet tastes, which appears to be instinctive (Steiner 1987), and the avoidance of bitter tastes would protect against the consumption of plant foods containing toxic alkaloids, or other bitter plant constituents. Sensory factors would therefore be very important in avoiding potentially dangerous foods. Learned associations between the unpleasant effects of consuming toxic or microbiologically contaminated foods would have been important for human survival (Garcia *et al.* 1974).

(*c*) *Satiety*

This is the power of foods to assuage hunger or the desire to eat. Although the mechanisms involved are unclear, there are strong physiological arguments for evoking a cascade of effects: a short term one that induces the response to stop feeding, a satiating effect and satiety, a long term one that prolongs the interval between periods of consuming food. The latter is strongly influenced by the fat content of a meal, and thus a learned association of fatty foods in delaying the need to seek more foods is possible (Rogers & Blundell 1990).

(*d*) *Social transmission*

Many of the learned associations between the consumption of certain foods are transferred by social interactions, especially during childhood when an infant's early exploration of the environment by oral means leads to training in which materials are acceptable as foods. In hunter–gatherer societies the children are exposed to the norms of that society very early in life.

Many of these societal norms develop into a complex set of food taboos and religious food laws which serve to protect the society from food-borne disease (Fieldhouse 1986). The food taboos may also restrict the exploitation of an available food resource, and introduce conservative food behaviours that may be detrimental, for example the avoidance of fish in some communities, and the prohibition of women eating eggs in others.

Food choice is an integration of psychosocial and biological influences (Rogers & Blundell 1990) and factors discussed up to now have not included the physiological determinants of food choice; early human societies had to select a diet which satisfied nutritional requirements.

4. SELECTION OF DIETARY MIXTURES TO MEET NUTRITIONAL REQUIREMENTS

In this section I shall discuss how the foods available to early human societies could be combined to meet their nutritional requirements. It is clear that suitably nutritious combinations were achieved but there has been substantial debate about the type of diet eaten during early human evolution (Eaton & Konner 1985). Others have argued for the benefits of the natural diet in relation to the effects of present-day 'unnatural' diets on the incidence of chronic disease (National Research Council 1989*a*). The argument revolves around the thesis that humans evolved under the influence of this natural diet and are therefore adapted to it.

Before we consider how the available foods could be combined to meet nutritional requirements it is

Table 5. *Estimates of energy requirements*

male	body mass 65 kg; height 1.72 m estimated basal metabolic rate (BMR) 284 kJ or 68 kcal h^{-1}.		
activity	time h	kJ	kcal
sleeping at 1 × BMR	8	2272	544
hunting at 3.6 × BMR[a]	8	8179	1954
sitting resting 1.4 × BMR	4	1590	381
sitting working 2.1 × BMR[b]	4	2386	571
total for 24 h:		14427	3454
female	body mass 50 kg; height 1.60 m estimated BMR 220 kJ or 53 kcal h^{-1}.		
activity	time h	kJ	kcal
sleeping at 1 × BMR	8	1760	424
gathering food and water[c] at 3.4 × BMR[c]	6	4488	1081
preparing food 2.7 × BMR	2	1188	286
other activities 2.0 × BMR	8	3520	848
total for 24 h		10956	2639

[a] Hunting pig.
[b] Carving weapons.
[c] Walking at normal pace.

important to establish estimates of physiological requirements. These are different from current dietary recommendations, for example the Recommended Dietary Allowances (RDA) issued by the National Research Council of the U.S.A. (National Research Council 1989*b*) and other official bodies, which include safety factors to meet the needs of the majority of the population.

The primary nutritional requirement for any human population is to satisfy energy needs as the proper utilization of other nutrients is dependent on adequate energy metabolism.

It is important to note that the primary physiological requirement, that of an adequate and regular supply of water, must also be met. Many present hunter–gatherers live in habitats where water is scarce and where the primary drive for survival is the securing of water. One imagines that before agricultural development the choice of habitat was less restricted, and settlement near water fulfilled this requirement and contributed to the food supply incidentally.

(*a*) *Energy requirements*

The estimation of energy requirements demands assumptions about body size and physical activity. By using the equations suggested by the Food and Agriculture Organisation, World Health Organisations and the United Nations University (1985) for estimating resting metabolic rates and the expenditures on hunting and gathering activities for males and females respectively, it is possible to arrive at reasonable estimates for use in constructing models for diet selection (see table 5).

These provide a basis for estimating the amounts of

Table 6. *Approximate weight of foods required to meet energy requirements*[a] *and protein supplied by that mass*

food	typical energy density/MJ kg^{-1}	mass to satisfy energy requirement/kg	typical protein content/g per kg	protein supplied/g
plant foods	—	—	—	—
fruits	3.68	3.4	7	24
nuts	16.33	0.77	132	101
seeds	14.65	0.85	110	94
seed legumes	2.97	4.21	63	265
leaves	1.21	10.3	21	217
roots and tubers	4.11	3.0	25	76
animal foods				
invertebrates				
insect grubs	12.9	0.97	137	132
insects	13.1	0.95	237	226
molluscs	3.49	3.6	136	487
crustacea	3.52	3.6	164	582
vertebrates	—	—	—	—
frog	2.85	4.4	153	671
lizard	6.89	1.8	275	499
snake	3.13	4.0	144	575
turtle	4.06	3.1	214	217
young birds	7.40	1.69	219	370
eggs	6.36	1.97	123	241
fish				
fatty fish	7.16	1.75	177	309
white fish	4.12	3.03	176	534
mammals				
rabbit	5.94	2.10	169	355
squirrel	4.60	2.71	263	715
pig	18.95	0.66	120	79
horse	7.13	1.75	190	333
buffalo	5.02	2.49	177	441

[a] This has been taken as 12.5 MJ per day per adult.

the different foods that would be required (see table 6). The calculations make no allowance for the amount of actual live weights of animals that would need to be caught.

A diet of leafy plant foods would require the greatest mass of food, and the amount of plant material (over 10 kg) is such that the bulk would be excessive both to gather and to consume. The advantages of nuts and seeds in terms of amounts to be collected are clear. The protein provided by the fruits is inadequate, and that by roots and tubers only marginally adequate.

The most effective food selection in terms of amounts therefore requires a proportion of animal foods, especially those that contain fat (Hayden 1981). These have several advantages; firstly, in the amounts required; secondly, as sources of substantial amounts of protein of high biological value, free from natural toxicants; thirdly, as sources of essential inorganic nutrients; fourthly, as sources of the B-vitamins, especially B12 and, very importantly, as a source of vitamin A and the other fat-soluble vitamins. Although the carotenoids can act as provitamin A these require an adequate fat intake for absorption.

(*b*) *Protein and other nutrients*

All the available foods, with the exception of fruits, would supply adequate protein and, provided energy needs were being met, protein supply would not be limiting. Effective food strategies in selecting a mixture of plant and animal foods would ensure that the biological quality of the protein would be satisfactory (FAO, WHO, UNU 1985).

All the foods that contain fat would provide essential fatty acids at levels above requirements, and the meats and fish would provide long-chain polyunsaturated fatty acids of the *n*-3 series (Paul & Southgate 1978).

The supply of inorganic nutrients from the plant foods alone would be marginally adequate, and the animal foods are important as sources of iron and zinc in biologically available forms. In the absence of dairy foods, calcium intakes would be low by western standards but biological adaptations to low calcium intakes are well documented, and exposure to sunlight would maintain high vitamin D status and maximize the efficiency of calcium utilization.

The supply of water-soluble vitamins, especially vitamin C and folates, from diets with a large amount of fresh plant foods would be more than adequate. The supply of fat-soluble vitamin A depends on animal sources, as does vitamin B12, and these provide circumstantial evidence for the role of animal foods in the early human diet. The animal foods available opportunistically are important potential sources of fat and micronutrients, and the importance of insects is very evident.

REFERENCES

Brand, J. C., Cherikoff, V. & Truswell, A. S. 1985 The nutritional composition of Australian Aboriginal Bush foods. 3. Seeds and nuts. *Fd Technol. Aust.* **37**, 279–297.

Brand, J. C., Rae, C., McDonnel, J., Lee, A., Cherikoff, V. & Truswell, A. S. 1983 The nutritional composition of Australian Aboriginal bush foods. 1. *Fd Technol. Aust.* **35**, 293–298.

Chambers, M. A. & Southgate, D. A. T. 1969 Nutritional study of the Islanders on Tristan da Cunha 1966: the foods eaten by Tristan Islanders, their methods of preparation and composition. *Br. J. Nutr.* **23**, 227–235.

Cherikoff, V., Brand, J. C. & Truswell, A. S. 1985 The nutritional composition of Australian Aboriginal bush foods. 2. Animal foods. *Fd Technol. Aust.* **37**, 208–211.

Coles-Rutishauser, I. H. E. 1986 Food intakes studies in Australian Aborigines: some methodological considerations. In *Proceedings of the* XII *International Congress of Nutrition* (*1985*) (ed. T. G. Taylor & N. K. Jenkins) pp. 706–710. London: John Libbey.

Crawford, M. A. 1968 Fatty acid ratios in free-living and domestic animals. *Lancet* i 1329–1333.

Crawford, M. A., Gale, M. M. & Woodford, M. H. 1978 Linoleic acid and linolenic acid elongation products in muscle tissue of *Syncerus caffer* and other ruminant species. *Biochem J.* **115**, 25–27.

Cummings, J. H., Southgate, D. A. T., Branch, W., Wiggens, H. S., Houston, H., Jenkins, D. J. A., Jivraj, T. & Hill, M. W. 1979 The digestion of pectin in the human gut, and its effect on calcium absorption on large bowel function. *Br. J. Nutr.* **41**, 477–485.

Eaton, B. S. & Konner, M. 1985 Paleolithic nutrition. *New Engl. J. Med.* **312**, 283–289.

Fieldhouse, P. 1986 *Food and nutrition: custom and culture.* London: Croom Helm.

Food and Agriculture and World Health Organisation United Nations University 1985 Energy and protein requirements. Report of an Expert Consultation of the FAO/WHO/UN. Geneva: WHO.

Garcia, J., Hankins, W. G. & Ruiswiak, K. H. 1974 Behaviour regulation of the milieu interne in man and rats. *Science, Wash.* **185**, 824–831.

Gonzalez, N. L. 1972 Changing patterns of North American Indians. In *Nutrition, growth and development of North American Indian children* (ed. W. M. Moore, M. M. Siberberg & M. S. Read) pp. 15–33. Washington: U.S. Dept Health and Welfare. Publication No NIH 72–26.

Hayden, B. 1981 Subsistence and ecological adaptations of modern hunter gatherers. In *Omnivorous primates gathering and hunting in human evolution* (ed. R. S. O. Harding & G. Teleki) pp. 344–421. New York: Columbia University Press.

Holland, B., Unwin, I. D. & Buss, D. H. 1989 Milk products and eggs. Cambridge: Royal Society of Chemistry.

Jolly, A. 1985 *The evolution of primate behaviour* pp. 45–71. New York: Macmillan.

Kuhnlein, H. V. 1989 Nutrient values in indigenous wild berries used by Nuxalk people of Bella Coola, British Columbia. *J. Fd Comp. Anal.* **2**, 28–36.

Kuhnlein, H. V. 1990 Nutrient values in indigenous wild plant greens and roots used by Nuxalk people of Bella Coola, British Columbia. *J. Fd Comp. Anal.* **3**, 38–46.

Le Gros Clark, F. 1967 Human food habits as determining economic and social life. In *Proceedings of the Seventh International Congress of Nutrition* vol. 4 (ed. J. Kühnau), pp. 18–24. Braunsweig: F. Vieweg & Sohn.

Liener, I. E. 1989 (ed.) *Toxic constituents of plant foods.* New York: Academic Press.

National Research Council 1989*a* *Diet and health; implications for reducing chronic disease risk.* Washington D.C.: National Academy Press.

National Research Council 1989*b* *Recommended dietary allowances* 10th edn. Washington D.C.: National Academy Press.

Paul, A. A. & Southgate, D. A. T. 1978 *McCance and*

Widdowson's the composition of foods 4th ed. London: H.M.S.O.

Pirie, N. W. 1973 Plants as sources of unconventional protein foods. In *The biological efficiency of protein production* (ed. J. G. W. Jones), pp. 101–118. Cambridge University Press.

Rogers, P. J. & Blundell, J. E. 1990 Physichobiological bases of food choice. *Nutr. Bull. Br. Nutr. Found.* **15**. (Suppl. 1) 31–40.

Schrire, C. 1984 (ed.) *Past and present in Hunter gather studies.* Orlando: Academic Press.

Selvendran, R. R. 1984 The plant cell wall as a source of dietary fibre: chemistry and structure. *Am. J. clin. Nutr.* **39**, 320–337.

Silverstone, G. A. 1985 Possible sources of food toxicants: plants, some foods of animal origin and food additives. In *Diet-related diseases* (ed. S. Seely, D. L. J. Freed, G. A. Silverstone & V. Rippere), pp. 40–118. Westport: Avi Publishing.

Southgate, D. A. T. 1984 Natural or unnatural foods. *Br. med. J.* **288**, 881–882.

Stahl, A. B. 1984 Hominid dietary selection before fire. *Curr. Anthropol.* **25**, 151–168.

Steiner, J. E. 1987 What the neonate can tell us about Umami. In *Umami: basic taste* (ed. Y. Kawamura & M. R. Kare), pp. 97–123. New York: Marcel Dekker.

Talbot, L. M. 1964 Wild animals as sources of food. In *Proceedings of the Sixth International Congress of Nutrition* (ed. D. P. Cuthbertson, C. F. Mills & R. Passmore), pp. 243–251. Edinburgh: Livingstone.

Wu Leung, W.-T. 1969 *Food composition table for use in Africa.* Rome: FAO.

Discussion

K. Hawkes (*Department of Anthropology, University of Utah, Salt Lake City, U.S.A.*). In the list of determinants of diet choice Professor Southgate did not include acquisition costs, something apart from availability. Some potential foods may be quite abundant in the environment of human foragers, and may even have quite high nutrient per unit mass values, yet be very costly to process. One view of the 'mesolithic transition' that happened in many places about 8000–18000 years ago emphasizes the incorporation of previously unexploited 'high cost' resources – like grass and tree seeds – into the diet as more profitable alternatives became increasingly scarce. These processing costs seem to play an important role in diet choice among contemporary human foragers; would Professor Southgate not expect this to be quite general?

D. A. T. Southgate. Yes, I was alluding to this in relation to the productivity of the food ranging area; clearly some foods would require considerable effort to collect. However, I suspect that sensory or satiety factors would still be more dominant in choice; satiety value is closely linked to energy density so that these factors are linked.

O. T. Oftedal (*Smithsonian Institution, Washington D.C., U.S.A.*). As, in most cases, animals cannot directly sense nutrients and must base their food preferences on sensory perceptions, and as animals in different environments must be faced with divergent sensory characteristics of foods, does Professor Southgate not think that evolutionary pressures may have moulded the perceptual abilities and preferences of different species in different ways? Is there evidence of differences in sensory abilities and preferences among different species?

D. A. T. Southgate. Quite clearly this is so, but I do not think that the basis of the differences in sensory preferences is understood. They may relate to differences in the metabolic processing of food constituents.

O. T. Oftedal. Professor Southgate mentioned that vitamin B_{12} must come from animal foods and yet clearly some primates eat little if any animal foods in the wild. Does he think microbial contamination or microbial activity on plant products could supply adequate vitamin B_{12}?

D. A. T. Southgate. It is possible that some humans obtain vitamin B_{12} from microbial contamination of foods or the environment, but I suspect that the small amounts of animal foods consumed by primates when foraging on leaves and fruits could also provide a source of vitamin B_{12}. Many reports of vitamin B_{12} in plants have resulted from contamination or analytical artifacts.

I. Crowe (*23 Lockhart Close, Dunstable, Bedfordshire, U.K.*). I believe that one of the factors affecting the choice of foods is visual information – to locate, identify, and to determine suitability – learnt by example, with colour vision being a prime factor: a characteristic of primate vision shared by the birds who originally exploited the same food resources, i.e. fruit.

D. A. T. Southgate. I agree: detection of fruits or leaves at the desired maturity would depend greatly on colour discrimination.

Contemporary human diets and their relation to health and growth: overview and conclusions

ELSIE M. WIDDOWSON

Formerly of The Department of Medicine, University of Cambridge, Cambridge CB2 2QQ, U.K.

SUMMARY

Contemporary human diets are probably as diverse now as they have ever been in the history of mankind. The abundance of food in the western world is in stark contrast to the lack of food and near starvation in parts of Africa, the continent where *Homo sapiens* evolved. The development of agriculture has enabled the population of the world to expand and to colonize almost the whole of its land surface, but the dependence on one staple food has introduced problems. If the staple crop fails, for example because of drought, there may be no alternative, and undernutrition and starvation are the result. Further, if the rest of the diet does not provide the nutrients that the staple food lacks, diseases due to specific nutrient deficiencies become widespread. Vitamin deficiencies among adults are less common now than they were 50 years ago, but even today millions of children in the poor rice-eating areas of the world are blind because their diets were deficient in vitamin A. For physiological reasons infants and young children will suffer most wherever there is a scarcity of food, of water, or of specific nutrients.

1. INTRODUCTION

One of the reasons for man's ability to colonize almost the whole of the world's land surface, including the colder regions, has been his adaptability in the matter of food. His omnivorous nature and his intelligence have enabled him to make use of plants and animals in his surroundings, and to grow and produce a wide variety of foods suitable to the environment in which he lives. He has developed methods of storage and transport so that food materials can be used all round the year and transported to all parts of the globe. The contemporary housewife in the developed countries can forage among the loaded shelves of the supermarket for out-of-season foods and foods from far off lands as well as those from her own, so that she can buy whatever she and her family desire. This is in stark contrast to the woman from drought-stricken Ethiopia searching for a few leaves or berries to help to assuage her own and her children's hunger for one more day. These are the extremes in contemporary food gathering and food consumption. Between them lies an immense range of diets, some nutritionally adequate, others limited in quantity or quality with regard to the foods available, the ability to grow them or the money to buy them and to the religious and other taboos that prevail. It is impossible to deal in detail with them all. This paper will therefore describe, in general terms, the types of diet eaten in different parts of the world today and their effect on health and growth. The extremes will be discussed in more detail, particularly those that are deficient in quantity and lead to chronic or acute undernutrition or starvation, and to diets deficient in quality, that is in specific nutrients, particularly protein and vitamin A. The vital importance of a supply of clean water will be emphasized. In conclusion some suggestions will be made as to how far what has been said about contemporary human diets and their relation to health and growth may have applied to primitive man, and may still apply to the other primates whose foraging habits have been described in previous papers.

2. THE IMPORTANCE OF THE STAPLE FOOD IN CONTEMPORARY HUMAN DIETS

In those parts of the world where agriculture in some form is practised and crops are cultivated, a plant, generally a cereal, has provided the staple food. In affluent countries where a wide range of animal and plant foods is available the staple food contributes less to the total diet than it does in areas where the supply of other foods is more limited in quantity and variety, and where the staple may provide a large part of the energy value of the diet. Here the nutrient composition of the staple food is of great importance, not only the amount of nutrients in the original cereal grain, but also the way in which the grain is treated before being eaten.

Table 1 shows the protein, fat, carbohydrate and thiamin (Vitamin B_1) in 100 g of the most widely used cereals in contemporary human diets. Wheat grows in temperate climates and is the cereal of Europe, North America, Australia, New Zealand and parts of Russia. It also grows at high altitudes in the tropics as in the northern parts of India. It is characterized by its high protein content, but this is largely a matter of plant breeding, and the protein content can vary from 8 to 14%. The values in the table apply to wheat used for

Table 1. *Composition of cereals used as staple foods*

(Data from Paul and Southgate (1978); Holland *et al.* (1988).)

	protein/g	fat/g	values per 100 g carbohydrate/g	energy/ kcal	thiamin/mg
wheat flour					
wholemeal	12.7	2.2	63.7	310	0.47
white	12.5	1.4	75.3	341	0.10
rice					
husked	6.7	2.8	81.3	357	0.40
polished	6.5	1.0	86.8	361	0.08
cornmeal (maize)					
whole unsifted	9.8	3.8	71.5	353	0.30
sifted	9.4	3.3	73.1	368	0.26
hominy	8.7	0.8	77.7	362	0.13
millet flour	5.8	1.7	75.8	354	0.68

bread-making where a high protein wheat is required. The effect of milling to produce white flour is to reduce the protein and fat a little because these are more concentrated in the outer layers and germ of the wheat grain than in the white endosperm, but the most important difference is in the thiamin, which is reduced by 80%. Because of this, and because people in Britain prefer white bread, all white flour has been by law enriched with thiamin for over 40 years.

Rice is the staple food of the people of eastern and southern Asia. It is a swamp plant and it grows in the wetter and warmer regions, particularly in the deltas of the great rivers. It contains less protein than wheat used for bread-making and more carbohydrate. A variety of methods are used for husking paddy, the whole rice grain. Hand-pounding is still practised in many parts of the East; the paddy is placed in a small stone or wooden mortar and pounded with a wooden pestle about six feet long. This breaks the outer husk which is removed by winnowing. Some of the germ and pericarp are also removed, but some are retained so that husked rice still contains most of the thiamin. However, mechanized rice mills have operated for many years and these produce a highly refined rice, polished rice, which contains very little thiamin. More of the vitamin may be lost by washing the rice after it has been bought in the bazaar. Thiamin is soluble in water, and is thrown away with the washing water.

There is another method of preparing the rice grain which retains much of the B vitamins, including thiamin, which is known as parboiling. Unhusked rice is first soaked in water and then steamed or boiled. This splits the outer husk, which is then easier to remove and it also increases the amount of thiamin in the part of the grain remaining after milling because parboiling geletinizes the starch in the endosperm and this prevents removal of the germ which contains much of the thiamin in the grain (Hinton 1948).

Enrichment of rice with thiamin was tried in the 1940s in the Philippines; this prevented beri-beri in people eating little else but rice, but the process presents so many technical difficulties that it has rarely been undertaken on a large scale (Food and Agriculture Organization 1954).

Beri-beri, caused by a deficiency of thiamin, was at one time widespread among the rice-eating people of the East (Vedder 1913), where most of the diet consisted of polished rice, with little animal or plant food such as beans which are a good source of the vitamin. The disease is far less common now, and has virtually disappeared from the prosperous eastern countries, Japan, Malaysia and Taiwan, and from the large cities such as Hong Kong, Bangkok and Rangoon. This is due largely to improvement in social and economic conditions and to the consumption of a better all-round diet so that rice makes a much smaller contribution than was the case 50 years ago (Passmore & Eastwood 1986).

Maize is the staple food in Central America where it originated, and in many parts of South America, Mexico and in South Africa. It requires less water than wheat or rice and it gives a high yield per acre.† Sifted maize meal contains about 85% of the original grain, but hominy is a flour made from the starchy endosperm and is used as a porridge in South Africa where it provides a large part of the energy of the diet among the poor black population. In Mexico hominy is made into flat cakes called tortillas.

Maize has two characteristics that differentiate it from other cereals. Yellow maize is the only cereal that is a source of vitamin A. It contains a mixture of carotenoids, all of which have pro vitamin A activity. However, the use of maize as the staple cereal has long been associated with pellagra, a disease caused by a deficiency of another member of the vitamin B complex, nicotinic acid. The reason for this puzzled scientists for many years, for maize was not more deficient in nicotinic acid than other cereals. It was discovered that the nicotinic acid in cereals is in a bound form, consisting of macromolecules, which cannot be converted to free nicotinic acid in the human body (Kodicek 1962; Mason *et al.* 1973). Nicotinic acid can be synthesized from the amino-acid tryptophan, but the major protein in maize, zein, is unlike the proteins of other cereals in that it contains very little tryptophan. Maize therefore has two disadvantages. Not only is the nicotinic acid in it unavailable, but a precursor of the vitamin is also deficient. There is one other point about nicotinic acid and maize. The bound form of nicotinic acid can be converted to the available

† 1 acre = 4046.9 m^2.

form by treatment with alkali (Kodicek *et al.* 1959). In Mexico where maize is used, not as a porridge but as a flat bread, the maize grains are softened in lime water and then ground into a wet dough before being cooked. This explains why pellagra does not occur in Mexico.

Enrichment of maize with nicotinic acid has been used as a method of prevention of pellagra, but the overall economic state of the population has improved; all forms of animal food are good sources of nicotinic acid and as soon as these form part of the diet the deficiency of maize becomes far less important.

Millets are resistant to drought and are grown in dry regions in Africa and in some parts of Asia and South America. They contain less protein than other cereals. The millet grains are de-husked and then soaked and boiled and made into a porridge or ground into a meal. This is done in the home, and refined millet products have not been produced on any large scale.

In some parts of Africa the staple food is not a cereal but a root (cassava or manioc) or a fruit (matoke, a cooking plantain or banana). These contain less protein than cereals, and protein provides only 3–4 % of the total energy instead of 8–16 % for cereals (Tan *et al.* 1985). Nutritional problems arising from their use will be discussed later (p. 293).

No staple food is by itself adequate for human well-being, whether of adults or still less of children. Cereals contain no vitamin C and only yellow maize contains carotenoids which are a source of vitamin A. They contain no vitamin D, but this may be unimportant to those living in the sunshine of tropical countries except among women who by custom keep themselves completely covered. Cereals also contain too little calcium and sodium for health and growth. Refined cereals are generally a poor source of B vitamins, and also iron. The remainder of the diet must provide these nutrients, and Southgate (this Symposium) has described the main sources of the nutrients that cereals lack. It is when these are not available in adequate amounts that the specific deficiency diseases arise.

3. CONTEMPORARY DIETS IN DEVELOPED COUNTRIES IN RELATION TO HEALTH AND GROWTH

In developed countries the general health of the population has improved and life expectancy has increased over the past 50 years. Children are growing more rapidly and they reach puberty earlier than they did before World War II (Tanner, 1973). This is due in part to better control of infections, especially those that affect young children, but better nutrition has also played a part. This is the positive side, but the abundance and free choice of foods have brought their problems. One is obesity. In most people, and in animals too, the appetite centres of the hypothalamus regulate the intake of energy to balance the expenditure, and this regulation is usually surprisingly exact. However, this is not always the case, and in some individuals the temptation of the sweet foods and foods with 'hidden' fat, so readily available, all highly calorific, overcomes the normal physiological responses of appetite, the individual takes in more energy than he or she expends and the result is obesity. This is characterized by a large excess of fat in cells of adipose tissue in the subcutaneous and deep body sites. Many studies have been made of the causes, effects and treatment of obesity (Royal College of Physicians 1983). One thing is certain; once a person has become obese it is extremely difficult for him, or more usually her, to return to a normal body mass and to maintain it (Garrow 1974).

Fat children tend to grow fast, in height as well as body mass, and they reach puberty earlier than their leaner counterparts. Their extra mass is not entirely due to fat, for they also have a greater mass of lean body tissue (Cheek *et al.* 1970; Forbes, 1987). Although obesity is disabling, and is a causative factor in some diseases, for example diabetes, it is not of itself likely to be a cause of death. In fact rapid deposition of fat in the body is physiological in some circumstances, for example during the suckling period in young mammals, and in preparation for periods of starvation during migration of birds, hibernation of animals, and during lactation in some species of marine mammals.

Several diseases whose incidence has increased over the past 50 years are believed to be associated with some aspect of the contemporary diet of abundance. One of these is cardiovascular disease. Conferences have been held and committees have met from time to time to consider the problem. High intakes of total fat, of fats with a large percentage of saturated fatty acids, of sugar, of cholesterol, of total food and energy and even of soft water have all been implicated as risk factors (Royal College of Physicians and British Cardiac Society 1976; World Health Organization 1982*a*; Department of Health and Social Security 1984). It now seems certain that no one food or nutrient can be singled out as the cause. The aetiology of cardiovascular disease is a complicated one and it involves influences other than food.

In 1969 Burkitt pointed out that cancer of the large intestine, which was common in western countries, was almost unknown in tropical Africa. Burkitt *et al.* (1972) suggested that the large amount of what was called dietary fibre in the largely vegetarian diets of the people in Central Africa prevented the disease by causing a more rapid transit of the contents of the large intestine. The passage of a larger bulk of more watery faeces exposed the intestinal mucosa for a shorter time to any oncogen that might be present than in a person eating a western-type diet with little dietary fibre and producing small amounts of more solid faeces. Burkitt's ideas and suggestions (Burkitt & Trowell 1975) led to a great deal of research into the components of dietary fibre and their role in the physiology and function of the large intestine. In 1980 The Royal College of Physicians published a report on the medical aspects of dietary fibre. It concluded that it was highly probable that increasing the proportion of dietary fibre in the diet of western countries would be nutritionally desirable.

Another disease that has become of concern over the past 50 years is osteoporosis. This is not directly related to diet, or some component of it, but it is more common now because we are living longer. From the age of 30 or 40 years the loss of bone on the inner surface through the activity of the osteoclasts exceeds the deposition of bone on the outer surfaces so that the shaft of the long bones becomes thinner, and the vertebrae also lose bone substance. Men have thicker bones than women, but women lose more bone in later life, particularly after the menopause, and their thinner bones are more liable to fracture after a fall. This loss of bone is part of the physiological process of ageing, and is accompanied by loss of protein, not only from bone but also from the soft tissues, particularly muscle, which loses potassium as well (Cohn *et al.* 1980; Forbes 1989; British Nutrition Foundation 1989). Decline in physical activity is undoubtedly one reason for the loss of hard and soft tissues, but a decrease in the secretion of anabolic hormones is thought also to be involved. Giving additional calcium to the elderly seems to have no beneficial effect (British Nutrition Foundation 1989). It has been suggested that a high intake of calcium during childhood might increase the amount of calcium in the bones and this would be an advantage when bone begins to be lost (Newton-John & Morgan 1968). This has yet not been put to the test.

4. UNDERNUTRITION AND STARVATION

Famines must always have occurred from time to time and there are many records of them in the Bible and other early writings (McCance & Widdowson 1951). They have generally been caused in the first instance by natural disasters, droughts or floods, with consequent failure of the staple crop, but they have frequently been made worse by wars and mass movements of the population. Sometimes deprivation of food has been deliberately imposed, as in the concentration camps of World War II, or self-imposed, as among the men in the Maze prison in Northern Ireland who starved themselves to death a few years ago. Starvation implies that little or no food is available, and is facing millions of people in Africa at the present time. Undernutrition arises when some food is eaten, but the amount is insufficient to meet physiological requirements. It may be chronic, as in parts of India, Africa and Latin America, or the shortage of food may occur suddenly among previously well-nourished people, as for example in The Netherlands during the railway strike in 1945, and in Germany after the end of the war.

In many ways the effects of starvation and undernutrition on the human body are similar and the differences are those of degree. Descriptions are given by Keys *et al.* (1945), Members of the Department of Experimental Medicine, Cambridge (1951), and Helweg-Larsen *et al.* (1952).

In adults there is a decline in body mass, but the organs and tissues lose their mass at different rates. The least essential tissues suffer first, the adipose tissue cells lose fat, which is oxidized to provide the energy necessary for the metabolic processes that must go on in the body if life is to be preserved. Some loss of protein is inevitable, and this comes in large part from skeletal muscle. The muscle fibres shrink, and in an undernourished person the spaces between them and between the empty adipose cells are occupied by an extracellular watery gel. In older persons some of the excess fluid collects in the legs and shows up as hunger oedema. In starvation, dehydration is more likely than overhydration. In undernutrition and starvation the skeleton retains its outward shape, but the fatty marrow disappears from the long bones and the cavity becomes filled with aqueous material. The skin becomes thinner and easily infected, and appears to hang loosely on the bones. The internal organs decline in mass, the liver sometimes a great deal. The heart and kidneys tend to lose mass in parallel with the body but the brain retains its size and structure. The alimentary tract becomes thin, and both mucosa and muscle are affected.

Despite all these structural changes the functions of the organs and systems of the body remain normal until the body has lost a great deal of its mass. The resting metabolic rate per kg of body mass is not greatly reduced in moderate undernutrition, but when the energy deficit is severe and there has been great decline in mass then the metabolic rate per kilogram falls, and also the body temperature. In moderate undernutrition the digestive tract functions normally provided there is no infection, but in starvation diarrhoea may become severe even when there is no intestinal infection, and the dehydration resulting from loss of fluid may prove fatal.

Undernourished children grow, but they grow slowly, and are underweight (wasted) and underheight (stunted) for their ages. Waterlow (1979) emphasized that such children suffer from mental and behavioural handicaps as well as nutritional, because the home where the undernutrition develops is also one where the child receives little or no mental stimulation.

Starving or near-starving children, particularly infants, who lose weight are in far worse plight than adults, for their requirements for energy per kilogram body mass are greater, as are their requirements for water. If the supply of breast milk fails, as it does if a woman is deprived of food for any length of time, the infant will be deprived of water and energy and will die before the adult. The lack of water is a more likely cause of death than lack of other nutrients.

Undernutrition reduces fertility in experimental and farm animals, but how undernourished a human population must be before fertility is reduced is a difficult question. Many of the populations that are increasing in numbers particularly rapidly are not well-nourished. Undernutrition of the mother retards the growth of the foetus, but not by a great deal. Chronically under-nourished women in developing countries often lactate well; this may be due to the custom of frequent breast-feeding which is often practised and which is known to increase the total volume of milk (Prentice 1980). In starvation, secretion of milk fails, and this is accentuated by lack of water.

5. PROTEIN DEFICIENCY AND KWASHIORKOR

In the 1950s and 1960s there was a great deal of interest in a disease of young children being due to a dietary deficiency, and recognized as such in 1932 by

Dr Cicely Williams, a medical officer in what was then the Gold Coast. She suggested that it was due to a deficiency of protein and used the local name for it, kwashiorkor. It was not until after the war, however, that scientific studies of the causes and manifestations of malnutrition among young children began to be made. The situation varied in detail in different countries; kwashiorkor was typically a disease of the weaned child, one or two years old, who was by tradition given the adult type of diet to eat. In Uganda, where kwashiorkor was prevalent, the staple food was matoke, the cooking banana, which provides only 4% of its energy as protein. The fruit is peeled and steamed and kneaded into a soft mash. The adults eat this with their fingers, dipping each mouthful into a sauce made from plants and some meat or fish. This provides them with additional protein. The newly weaned child has the same food available but is not helped over eating. It manages to eat the matoke but is not able to manipulate this into the sauce. It needs more protein in relation to energy than the adult, but it gets less. The abundant carbohydrate in the matoke at first exerts its well-known protein-sparing effect, but ultimately the characteristic symptoms of protein deficiency appear. The child, often quite fat, becomes grossly oedematous with low serum proteins and other biochemical abnormalities (Trowell *et al.* 1954).

Such children lose their appetites and become utterly miserable. Sometimes they are force-fed with a high carbohydrate food, but this only makes matters worse.

As time went on it was realized that kwashiorkor, due primarily to a deficiency of protein, was one end of a spectrum of malnutrition of the infant and young child, and that marasmus, or near starvation, due primarily to a lack of food and energy, was the other. Between them lay a whole range of deficiencies of both energy and protein, and in any case a marasmic child is short of protein because it is short of food and a child with kwashiokor is short of energy because it will not eat. This whole spectrum of infantile and childhood malnutrition is now covered by the term 'protein–energy malnutrition'.

In 1974 McLaren wrote an article entitled 'The great protein fiasco' in which he set out his views that far too much attention had been paid by relief organizations to protein deficiency among children, so that it was believed that only foods rich in protein would be of any use. The world-wide problem was at the other end of the spectrum, energy deficiency and marasmus. This is even more true today than it was in 1974.

6. VITAMIN A DEFICIENCY, XEROPHTHALMIA AND BLINDNESS

It has been estimated that at the present time three million children under ten years of age are blind because their diets were deficient in vitamin A, and 20–40 million are less severely affected (WHO 1991). It is mainly in the poor rice-eating areas where vitamin A deficiency is most common (McLaren 1986*a*). About half the affected children are in the Indian sub-continent, the others in Africa, South and Central America, near Eastern countries, and China, Burma and the Philippines. Young children are particularly at risk because the growth rate is then most rapid, and vitamin A requirements are closely related to growth. Infections, which are still common among children in these countries, also increase the requirement for vitamin A (McLaren 1986*b*).

Vitamin A, or retinol, is essential for the normal function of the eye, both of the retina and the epithelial tissues of the cornea. In the retina an aldehyde of retinol is a component of the pigment rhodopsin, which is involved in the visual cycle concerned with night vision. In vitamin A deficiency there is insufficient rhodopsin in the retina, and this leads to night blindness. More serious is the effect of a deficiency on the cornea. The cells of the conjunctiva covering the cornea become keratinized; this ultimately leads to a softening and destruction of the cornea, xerophthalmia, and to total blindness (World Health Organization 1982*b*). Once a child has developed xerophthalmia the blindness cannot be cured (World Health Organization 1991). The foods that contain most vitamin A are milk fat, egg yolk, liver and some fatty fish. Carotene, the precursor of vitamin A, is found chiefly in yellow vegetables and green leaves in association with chlorophyll, so that dark green leaves are a good source of carotene, while pale green leaves are not. The only vegetable oil rich in carotene is red palm oil, produced in West Africa and Malaysia. In those countries were vitamin A deficiency occurs young children do not get appreciable amounts of any of these foods. They come from poor sections of the community, they are generally undernourished, and often suffer from various infections. Ignorance is partly to blame; dark green leaves which are usually available, are despised as 'poor man's food' (Pirie 1983).

7. THE VITAL IMPORTANCE OF WATER

Without food an adult can survive for 70 days, but without water in a temperate climate he can only survive for 7–8 days. In a hot environment, when losses of water by the skin are increased, survival time without water is correspondingly less. An infant has a larger turnover of water than an adult. It loses more by the lungs because it breathes a greater volume of air per kg body mass each minute, and more by the skin because it has a larger surface area in proportion to its mass. It also requires a larger volume of water to excrete urea and salts by the kidneys because it cannot concentrate urine to the same extent (McCance 1948). A young infant normally takes about 10% of its body mass of water each day, which would correspond to 7 l for a 70 kg man. An infant, therefore, will die of dehydration sooner than an adult if deprived of water.

If no clean water is available, as was the case recently among the Kurds who escaped to the mountains and the Bangladeshi people after floods, men, women and children will drink contaminated water rather than none at all. The result is intestinal infection and diarrhoea, sometimes very severe, which accentuates the dehydration. Deaths of children in these situations are inevitable.

8. OVERVIEW AND CONCLUSIONS

In this survey of contemporary human diets, most attention has been paid to excesses and deficiencies, rather than to satisfactory mixtures of foods in optimal amounts, so that those who consume them remain healthy to a ripe old age and their children grow well in height and body mass. The range of foods that can make up a satisfactory human diet is vast. If foods of animal origin are included the diet is more likely to provide all the nutrients required, but a completely vegetarian diet can supply all that is needed except vitamin B_{12} which is contained only in animal foods. The papers presented at this meeting show how the higher primates, including the human species, are still able to forage for a mixture of foods which provide all the nutrients they require. Serious deficiencies of vitamins and other nutrients do not appear to occur, and from the descriptions of the diets they are hardly to be expected, and nor are troubles associated with an excess of energy or a lack of dietary fibre. If food supplies become scarce, however, as they may do if habitats are lost, or water supplies are inadequate, then monkeys and apes, and hunter–gatherers too, will suffer effects of undernutrition and starvation similar to those that have been described. It is to be expected that young animals, like human infants, will be particularly vulnerable, as they will be also if there is not an adequate supply of water. Behaviour is important, for when human beings and animals are hungry they will struggle and even fight for food. This was evident among the Kurdish refugees when relief supplies arrived; women were left behind and children were trampled on as men and older boys pushed forward to get the food. Cheney *et al.* (1988) refer to high mortality among low-ranking female vervet monkeys during periods of food scarcity, which suggests that they also suffered in their endeavour to get food.

The introduction of agriculture and the raising of animals and plants for food introduced a whole new dimension into human nutrition and enabled the world population to expand as it could never have done with a hunter–gatherer way of life. It brought problems however; research has identified the causes, but prevention depends among other measures on education, greater prosperity, freedom from war, and birth control, and these are outside the scope of this meeting.

REFERENCES

British Nutrition Foundation 1989 *Calcium. The Report of the British Nutrition Foundation's Task Force.* London: The British Nutrition Foundation.

Burkitt, D. P. 1969 Related disease – related cause? *Lancet* ii. 1229–1231.

Burkitt, D. P. & Trowell, H. C. 1975 *Refined carbohydrate foods and disease. Some implications of dietary fibre.* London: Academic Press.

Burkitt, D. P., Walker, A. R. P. & Palmer, N. S. 1972 Effect of dietary fibre on stools and transit times, and its role in the causation of disease. *Lancet* ii. 1408–1412.

Cheek, D. B., Schultz, R. B. & Parva, A. 1970 Overgrowth of lean and adipose tissue in adolescent obesity. *Pediat. Res.* **4**, 268–279.

Cheney, D. L., Seyfarth, R. M., Andelman, S. J. & Lee, P. C. 1988 Reproductive success in vervet monkeys. In *Reproductive Success* (ed. T. H. Clutton-Brock), pp. 384–402. Chicago University Press.

Cohn, S. H., Vartsky, D., Yashimura, S., Sevitsky, A., Zanzi, I., Vaswani, A. & Ellis, K. J. 1980 Compartmental body composition based on total body nitrogen, potassium and calcium. *Am. J. Physiol.* **239**, E524–E530.

Department of Health and Social Security 1984 *Diet and cardiovascular disease* (*Rep. Health. Subj., Lond.* No. 28.). London: HMSO.

Food and Agriculture Organization, 1984 *Rice enrichment in the Philippines* (Nutritional Studies, No. 12). Rome: FAO.

Forbes, G. B. 1987 *Human body composition.* New York: Springer Verlag.

Garrow, J. S. 1974 *Energy balance and obesity in man.* Amsterdam: North-Holland Publishing Company.

Helweg-Larsen, P., Hoffmeyer, H., Kieler, J., Hess Thaysen, E., Hess Thaysen, J., Thygesen, P. & Wulff, M. H. 1952 *Famine disease in German Concentration Camps. Complications and sequels.* Copenhagen: Ejnar Munksgaard.

Hinton, J. J. C. 1948 Parboiling treatment of rice. *Nature, Lond.* **162**, 913–915.

Holland, B., Unwin, I. D. & Buss, D. H. 1988 *Cereals and cereal products* (the third supplement to McCance & Widdowson's *The Composition of Foods*, 4th edn. The Royal Society of Chemistry and Ministry of Agriculture, Fisheries and Foods). Old Woking, Surrey: Unwin Bros.

Keys, A., Brozek, J., Henschel, A., Mickelsen, O. & Taylor, H. L. 1950 The biology of human starvation, vols I and II. Minneapolis: The University of Minnesota Press.

Kodicek, E. 1962 Nicotinic acid and the pellagra problem. *Bibl. Nutritio Dieta* **4**, 109–127.

Kodicek, E., Braude, R., Kon, S. K. & Mitchell, K. G. 1959 The availability to pigs of nicotinic acid in *tortilla* baked from maize treated with lime-water. *Brit. J. Nut.* **13**, 363–384.

McCance, R. A. 1948 Renal function in early life. *Physiol. Rev.* **28**, 331–348.

McCance, R. A. & Widdowson, E. M. 1951 Famine. *Postgrad. med. J.* **27**, 268–277.

McLaren, D. S. 1974 The great protein fiasco. *Lancet* ii, 93–96.

McLaren, D. S. 1986*a* Pathogenesis of vitamin A deficiency. In *Vitamin A deficiency and its control* (ed. J. C. Bavernfeind), pp. 153–176. New York: Academic Press.

McLaren, D. S. 1986*b* Global occurrence of vitamin A deficiency. In *Vitamin A deficiency and its control.* (ed. J. C. Bavernfeind), pp. 1–18. New York: Academic Press.

Mason, J. B., Gibson, N. & Kodicek, E. 1973 The chemical nature of the bound nicotinic acid of wheat bran. Studies of nicotinic acid containing macromolecules. *Brit. J. Nut.* **30**, 297–311.

Members of the Department of Experimental Medicine, Cambridge 1951 *Studies of undernutrition, Wuppertal 1946–9* (*Spec. Rep. Ser. Med. Res. Coun.*, No. 275). London: HMSO.

Newton-John, H. & Morgan, D. B. 1968 Osteoporosis: disease of senescence? *Lancet* i, 232–233.

Passmore, R. & Eastwood, M. D. 1986 Human Nutrition and Dietetics, pp. 311–317. London: Churchill Livingstone.

Paul, A. A. & Southgate, D. A. T. 1978 *McCance and Widdowson's The Composition of Foods* (Fourth revised and extended edition of MRC Special Report, No. 297). London: HMSO.

Pirie, A. 1983 Vitamin A deficiency and child blindness in the developing world. *Proc. Nut. Soc.* **42**, 53–64.

Prentice, A. M. 1980 Variations in maternal dietary intake, birthweight and breast-milk output in The Gambia. In

Maternal nutrition during pregnancy and lactation (ed. H. Aebi & R. G. Whitehead), pp. 167–183. Bern: Hans Huber.
Royal College of Physicians 1980 *Medical aspects of dietary fibre*. London: Pitman Medical.
Royal College of Physicians 1983 Obesity. *J. R. Coll. Physicians* **17**, 5–65.
Royal College of Physicians and British Cardiac Society 1976 Prevention of Coronary Heart Disease. *J. Roy. Coll. Physicians* **10**, 213–275.
Tan, S. P., Wenlock, R. W. & Buss, D. H. 1985 *Immigrant foods* (second supplement to McCance and Widdowson's *The Composition of Foods*). London: HMSO.
Tanner, J. M. 1973 Trend towards earlier menarche in London, Oslo, Copenhagen, The Netherlands and Hungary. *Nature, Lond.* **243**, 95.
Trowell, H. C., Davies, J. N. P. & Dean, R. F. A. 1954 *Kwashiorkor*. London: Edward Arnold.
Vedder, E. B. 1913 *Beri-beri*. London: Bale and Davidson.
Waterlow, J. C. 1979 Childhood malnutrition – the global problem. *Proc. Nut. Soc.* **38**, 1–9.
Williams, C. 1931–32 Gold Coast Annual Report Med. Dept. Appendix E. Government Printing Office, Accra.
World Health Organization, 1982*a* *Prevention of coronary heart disease* (WHO Technical Report, Series No. 678). Geneva: WHO.
World Health Organization, 1982*b* *Control of vitamin A deficiency and xerophthalmia* (WHO Technical Report, Series No. 692). Geneva: WHO.
World Health Organization 1992 (In the press).

Discussion

O. T. OFTEDAL (*Smithsonian Institution, Washington, D.C., U.S.A.*). It is commonly stated that there are no protein stores as such because proteins in the body invariably perform functions. Yet some animals, such as seals, may mobilize large amounts of protein from tissues to support milk production for their young without apparent adverse consequences. If protein can be lost from the body without adverse effect, is it not, for all practical purposes, part of a 'store'?

E. M. WIDDOWSON. This is a matter of semantics. I presume that the female seal deposits protein in her body during pregnancy just as the rat does and probably also the human female. In the rat the muscles provide the 'store'. This is discussed in detail by Naismith (1981).

Reference

Naismith, D. J. 1981 Diet during pregnancy: a rationale for prescription. In *Maternal nutrition in pregnancy. Eating for two?* (ed. J. Dobbing), pp. 21–32. New York: Academic Press.

A. WHITEN (*Scottish Primate Research Group, University of St Andrews, U.K.*). Dr Widdowson notes that osteoporosis is more common in older people. Might not the increased occurrence of osteoporosis in the settled neolithic people studied by Dr Ulijaszek be the result of that population including a higher number of subjects living to greater ages?

S. J. ULIJASZEK (*Department of Biological Anthropology, University of Cambridge, U.K.*). This is difficult to answer, as the assessment of chronological age in skeletal material is prone to large errors above the age of about twenty years, and virtually impossible above the age of forty years. So it is impossible to say whether a greater proportion of people in Neolithic times lived longer than people in pre-Neolithic times. With low life expectancy at birth at both times, it is unlikely that a substantial proportion of the population at either time would have achieved ages at which osteoporosis could be attributed to ageing.

E. M. WIDDOWSON. It may be that the diets of the children of the Neolithic people did not contain sufficient bio-available calcium for the optimal growth and calcification of their bones or perhaps Neolithic women's menopause came at an earlier age than it does today.

Index

INDEX